THE
RIDDLER

FANTASTIC PUZZLES FROM
FiveThirtyEight

Edited by
OLIVER ROEDER

Foreword by
NATE SILVER

W. W. NORTON & COMPANY

INDEPENDENT PUBLISHERS SINCE 1923

NEW YORK · LONDON

For information about permission to reproduce selections
from this book, write to Permissions, W. W. Norton & Company, Inc.,
500 Fifth Avenue, New York, NY 10110

For information about special discounts for bulk purchases,
please contact W. W. Norton Special Sales at
specialsales@wwnorton.com or 800-233-4830

Manufacturing by: LSC Communications Harrisonburg
Book design by: MarySarah Quinn
Production manager: Julia Druskin

Library of Congress Cataloging-in-Publication Data

Names: Roeder, Oliver, editor. | Silver, Nate, 1978- writer
of foreword.
Title: The riddler: fantastic puzzles from
FiveThirtyEight / edited by
Oliver Roeder; foreword by Nate Silver.
Description: First edition. | New York: W. W. Norton & Company,
[2018] | Includes bibliographical references.
Identifiers: LCCN 2018027399 | ISBN 9780393609912 (hardcover)
Subjects: LCSH: Mathematical recreations. | LCGFT: Logic
puzzles. | Puzzles and games.
Classification: LCC QA95.R52 2018 | DDC 793.74—dc23
LC record available at https://lccn.loc.gov/2018027399

W. W. Norton & Company, Inc.
500 Fifth Avenue, New York, NY 10110
www.wwnorton.com

W. W. Norton & Company Ltd.
15 Carlisle Street, London W1D 3BS

1 2 3 4 5 6 7 8 9 0

THE RIDDLER

FOR

JACK TABOR—

farmer, baker, winemaker, games player, Riddler forebear

CONTENTS

LOGIC

PROBABILITY

GEOMETRY

FOREWORD

NATE SILVER

Because I run a website (FiveThirtyEight) that's devoted to statistical thinking, people assume I'm good at math.

The truth is, I'm not all that good at math. I still have nightmares about calculating derivatives, for example. I'm just a very determined problem solver. And I learned enough math along the way to solve some of the problems I was interested in.

When I was a kid, that included some (extremely dorky) problems, such as how to plan the optimal road trip for my family or how to beat adults at fantasy baseball. Like Billy Beane in *Moneyball*, I taught myself sabermetrics (the statistical study of baseball) not because of an interest in statistics per se but because I was tired of losing to everyone else.

Now I focus on problems like how to estimate the probability that one candidate will win a presidential election. That's a tricky problem because polls (ahem) aren't always perfectly accurate, and the Electoral College is mathematically complex. (The big issue is that what happens in one state—Wisconsin, say—is correlated with what happens in similar states, such as Michigan). But by using techniques like matrix algebra, and by being comfortable with probabilistic thinking ("the Republican has a 7 in 10 chance of winning"), it's possible to come up with some reasonable estimates.

It's this same problem-solving attitude that has made Oliver Roeder and The Riddler so successful. A few of the problems in this book could come straight from an algebra or a trigonometry textbook. But most of them require something else that isn't always taught in school: *mathematical intuition.*

These aren't trick questions designed to mislead you. Nor are they abstract and formal. Instead, they're enticingly thorny problems that resemble the ones that I and other applied statisticians encounter in the real world. Some of them have exact answers, but many don't, instead requiring you to think probabilistically. Even if your mathematical intuition is fairly strong, you may need to approach them in two or three different ways before you really start making progress.

For people like you and me and the rest of Oliver's readers, that might sound both fun and challenging. It's also something else: good practice. Mathematical intuition isn't necessarily something that most of us are born with. Nor is it something easily taught. *But the good news is that mathematical intuition can be learned.* It can be learned by application: by devoting ourselves toward problems we're interested in, like the problems in this book. I'm proud to present Oliver and The Riddler in book form, and hope you'll enjoy it as much as I do.

INTRODUCTION

In 1865, the British Museum received a brittle and breaking scroll of papyrus. The document was discovered in Egypt in the ruins of the mortuary temple of Pharaoh Ramesses II. It was well over 3,000 years old and worse for the wear. At the museum, curators carefully unrolled its two sides and placed them in two glass frames. One side had a jagged gash and the other had a blank section about 10 feet long. Yet despite its rough condition, much of the document could still be painstakingly deciphered and translated, and it became known as the Rhind papyrus, after the antiquarian who had purchased it. Its text begins: "The entrance into the knowledge of all existing things and all obscure secrets."

What followed turned out to be the world's oldest collection of math puzzles.

The problems posed in the Rhind papyrus won't present much of a challenge to the modern reader—especially for those of you holding this book. One of its problems, for example, reads, "Find the volume of a cylindrical granary of diameter 9 and height 10." Another: "Sum the geometrical progression of five terms, of which the first term is 7 and the multiplier 7." But now they serve a grander purpose. The 84 problems and solutions recorded on the papyrus provide some of the clearest insights into the numerical methods of the ancient Egyptians, some of the world's earliest mathematicians.*

In December 2015—150 years after the British Museum's acquisition and three millennia after the death of Ramesses II— the empirical journalism site FiveThirtyEight and I began publishing a weekly math puzzle column called "The Riddler." FiveThirtyEight, named for the number of electors in the American Electoral College, was born of a deep, abiding belief that sober data analysis and statistical modeling could add value to the frenzied political conversation. It's since taken that same approach to sports, science, economics, and culture. The answers it has provided have resonated with the site's readers.

With "The Riddler," we've gotten the chance to turn that conversation around. This time the readers provided the answers.

The puzzles and solutions in this book originated not from a dutiful ancient scribe but often from people like you. Each week, when the column is published, Riddler readers take to the far-flung boroughs of the internet: Twitter, Facebook, GitHub, Reddit, Stack Overflow, and forums beyond. There the column is dissected, discussed, and solved. My inbox then overflows with formulas, conjectures, diagrams, and videos. Awash in a week's worth of clever and surprising mathematical ideas, I publish as many as I can, post a new problem, and the work starts all over again. It's a testament not only to the great strides mathematics has taken in these past 3,000 years but also to the strength of technology to accelerate, combine, and disseminate ideas. Martin Gardner, who wrote a legendary

---

* As readers of this introduction have no doubt already calculated, the volume is about 636.17 cubic units and the sum is 19,607.

math column for *Scientific American* for many pre-internet years, wrote in his autobiography: "One of the pleasures of writing the column was that it introduced me to so many top mathematicians, which of course I was not. Their contributions to my column were far superior to anything I could write and were a major reason for the column's growing popularity."* That's precisely how I feel about the readers who have contributed to my column.

Now you are holding a physical testament to that collaboration. Within this book are 49 puzzles and solutions selected to appeal to a breadth of mathematical interests and a depth of mathematical skills. The simplest require a mere flash of logical insight. Others draw on the tools of trigonometry, geometry, combinatorics—and even a bit calculus. And the toughest involve deep applications of analysis and probability theory.

All of them are meant to be fun. In the ancient papyrus, the puzzles concerned the arithmetic of the practical: granaries, flour, beer, bread. The puzzles here go a bit further afield. Each has a story—perhaps set in a dystopian city, a playground at recess, or an NBA arena, to name a few (although you will also find two puzzles about pizza, without which no puzzle book is complete).

A quick note about the organization: As I culled the published columns and assembled new ones for this book, they tended to fall into three broad mathematical categories: logic, probability, and geometry. In the first category, you might find yourself rigging an election, outwitting a car salesman, or competing in a space race. In the second you may be fending off an

---

* Martin Gardner, *Undiluted Hocus-Pocus: The Autobiography of Martin Gardner* (Princeton, NJ: Princeton University Press, 2013), 136.

alien invasion, visiting a national park, or teaching your baby to walk. And in the third you might be baking a cake, outrunning an angry ram, or fighting over pizza with your siblings. While the boundaries are inevitably fuzzy, the categories should provide a rough map to where in the mathematical world you can expect to find yourself. Within each category, the problems tend to get more difficult the further along you go.

While there may be right and wrong answers to the puzzles, there is no right or wrong way to proceed through this book. If you are presented with a problem about pizza, say, perhaps you'd do well to pull out a pencil and paper and just get to it. Or maybe you'd be more comfortable on your computer, running some pizza simulations. Or maybe you find a hands-on approach is best, and you order up some actual pizzas to test your hypotheses. In any case, please enjoy them and continue this digital collaboration in the physical world.

OLIVER ROEDER

Brooklyn, New York

LOGIC

*Logic takes care of itself; all
we have to do is to look and
see how it does it.*

—LUDWIG WITTGENSTEIN

WHAT IF ROBOTS
CUT YOUR PIZZA?

At RoboPizza™ pies are cut by robots. When making each cut, a robot will randomly (and independently) pick two points on a pizza's circumference and then cut along the chord connecting them. If you order a pizza and specify that you want the robot to make exactly three cuts, what is the expected number of pieces your pie will have?

Submitted by Zach Wissner-Gross

SOLUTION:

You can expect your pizza to have five pieces, on average.

A key to solving this problem lies in realizing that the exact positions of the points the robot chooses when making its cuts are irrelevant. All that matters is whether the lines it slices—the pizza circle's chord lines—cross. (The exact positions of the points will affect the slice sizes, but here we only care about the number of slices.)

Imagine the robot picks six random points before making its required three cuts between them. It would then randomly pick any one point to start with, randomly choose one of the remaining five, and slice between them. Then it would pick another point to start with, pair it randomly with one of the three remaining points, and slice between those. Finally, the robot makes one last cut between the two remaining points. That's 15 possibilities. Five of these yield four pieces, six yield five pieces, three yield six pieces, and one yields seven pieces. Here is an illustration of the possible arrangements and their slice counts:

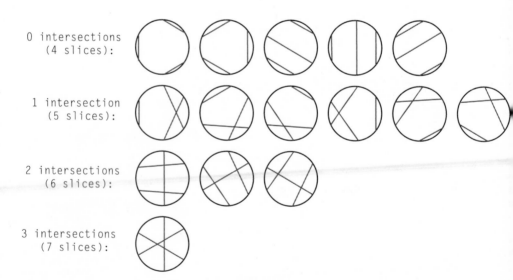

0 intersections (4 slices):

1 intersection (5 slices):

2 intersections (6 slices):

3 intersections (7 slices):

Thanks to the way the robot randomizes its cuts, each preceding scenario occurs with equal probability. Therefore, the average number of pieces is $(5 \times 4 + 6 \times 5 + 3 \times 6 + 1 \times 7)/15 = 75/15 = 5$.

(Note also some weird, knife-edge possibilities we didn't consider, such as the robot choosing the same chord twice or all three chords intersecting at one point, as you would normally try to slice a pizza. However, because these happen with probability zero—they are single possibilities out of infinite possible pizzas—they don't factor into our calculation of the expectation.)

There is also an elegant generalized solution: If the robot makes k random cuts, the expected number of slices is $(k + 2)(k + 3)/6$. This can be shown by induction: If the expected number of pieces resulting from k cuts is denoted by $E(k)$, then the expected number of pieces for $k + 1$ cuts is $E(k + 1) = E(k) + 1 + k/3$. Why? Slicing along a new random chord will always add at least one slice, and the new chord has a one-third chance of intersecting each of the already sliced chords. (Why one-third? The two points that define the added chord can either fall both on one side of an existing cut, both on the other side, or one on each side—in which case an intersection occurs. These three possibilities are equally likely.) We also know that one cut always gives two pieces; that is, $E(1) = 2$. Let's list a few terms.

$E(1) = 2$	$= 12/6$
$E(2) = 2 + 1 + 1/3 = 10/3$	$= 20/6$
$E(3) = 10/3 + 1 + 2/3 = 5$	$= 30/6$
$E(4) = 5 + 1 + 1 = 7$	$= 42/6$
$E(5) = 7 + 1 + 4/3 = 28/3$	$= 56/6$

We can now quickly see the general slice formula for k cuts working: $(k + 2)(k + 3)/6$.

Mmm, robopizza.

CAN YOU BEST THE MYSTERIOUS MAN IN THE TRENCH COAT?

A man in a trench coat approaches you and pulls an envelope from his pocket. He tells you that it contains a sum of money in bills, anywhere from $1 up to $1,000. He says that if you can guess the exact amount, you can keep the money. After each of your guesses, he will tell you if your guess is too high or too low. But! You only get nine tries. What should your first guess be to maximize your expected winnings?

Submitted by Dan Oberste

SOLUTION:

When given only nine chances to guess how much money is in the mysterious man's envelope, your first guess should be $745.

There are two key ideas here: (1) If you have one guess, you can only be sure of winning if you have one option left. Two guesses will allow you to win with three options: Pick the middle amount of the three and then, if it's wrong, the man telling you "higher" or "lower" will guarantee you can win with your last guess. (2) If you are making your final guess and are unsure which of multiple options is correct, you should always choose the highest one because it has a higher expected value. (It has the same chance of being right as other options, and it gives you more money if it's right.)

In general, using the strategy above, if you can win M options with N guesses, you can win $2M + 1$ options with $N + 1$ guesses. (You can win one option with one guess, you can win three options with two guesses, you can win seven options with three guesses, and so on.) Because you more or less split the options in half with each guess, you can win a $(2^n - 1)$-option game with n guesses. Thus, with your nine guesses, you can win a 511-option game with certainty.

Unfortunately, you have 1,000 options. But generalizing Idea 2, it is best to make all your ensured wins as large as possible. So pretend the numbers from 1 to 489 don't exist, and use Idea 1 on the numbers from 490 to 1,000. You'll lose 489/1,000 of the time, but that was going to happen anyway, and this strategy ensures that the situations in which you lose are the ones in which there was the least to gain. Thus your first guess should be 745 (1,000 minus 255, which is smack in the middle of your 511 options), and your expected winnings will be $380.695 (511/1,000 × 745, because 745 is the average value of the amounts you might win).

THE PERPLEXING PUZZLE
OF THE
PRIDEFUL PARTYGOERS

It's Friday and that means it's party time! A group of N people are in attendance at your shindig, some of whom are friends with each other. (Let's assume friendship is symmetric—if person A is friends with person B, then B is friends with A.) Suppose that everyone has at least one friend at the party and that a person is called "proud" if her number of friends is larger than the average number of friends that her own friends have. (A competitive lot, your guests.)

For example, let's say Anne, Bob, and Charlie are at the party. If Anne is friends with Bob and Charlie, but Bob and Charlie are not friends with each other, then Anne is "proud." She has two friends, while her friends have on average just one friend.

So how many of these cutthroat partygoers will feel chuffed around the punchbowl? In other words, how large can the share of "proud" people at the party be?

Submitted by Dominic van der Zypen

SOLUTION:

To maximize the number of proud people at a party (i.e., the number of people who have more friends than their friends do, on average), first imagine a situation where everyone at the party has an enormous number of friends. That is, everyone is friends with everyone else. In mathematical jargon, you would have a "complete graph"—a system of points where every point is connected to every other point. But if 100 percent of people are friends with each other, then no one has more friends than anyone else; thus no one would be proud.

However, if your party is the same as above except that exactly two people are not friends with each other, then the entire party is proud except for this unfortunate pair. (This would be a complete graph with one edge removed.) If $N > 3$ (and, really, you need at least three to party), then the largest share of proud people at the party is $(N - 2)/N$. In other words, if there are three people at the party, then one out of three can be proud. If there are 100 people at the party, then 98 of them can be proud. As the party expands toward infinity, the portion approaches 100 percent.

WHERE DID NIGEL GO?

Nigel, your eccentric friend who lives in England, tells you about his recent American holiday this way:

> I flew nonstop from Heathrow to an airport somewhere in the contiguous 48 states. I hired a car at the airport and spent two weeks driving around the country. I traveled the entire way by car, and I never put the car on a boat or plane. I stayed within the contiguous 48 states the entire time. I crossed the Ohio River exactly once, the Missouri River exactly twice, the Mississippi River exactly three times, and the Continental Divide exactly four times. I drove all the way to the Pacific and Atlantic Oceans, as well as to the Gulf of Mexico. At the end of my holiday, I returned the car to the same airport where I had hired it, and I then flew nonstop back to Heathrow.

What is the only U.S. state that you can say for sure Nigel visited during his recent American holiday?

Submitted by Dave Moran

SOLUTION:

Nigel definitely visited Minnesota. To cross a river an odd number of times and still end up where he started (at the same airport), Nigel must have "gone around" the river. With the Ohio River, that can be done any number of ways—for example, one can cross the river from Indiana to Kentucky or at a point where the river is entirely within Pennsylvania and then go around the mouth of the Ohio River by crossing the Mississippi River anywhere downstream of that point (near Cairo, Illinois) and then crossing the Mississippi again anywhere upstream of that point.

But to get around the Mississippi is harder. Since Nigel didn't put his car on a boat or plane, he must have gone around the source of the Mississippi, which is in Minnesota (at Lake Itasca). That means he had to actually visit Minnesota because avoiding Minnesota and going around the source would require a trip into Canada. Note that it is possible to go around the source in Minnesota without crossing into Wisconsin by crossing the river before or after going around Lake Itasca while the river is still in Minnesota. So Minnesota is the only state that Nigel visited for sure.

WOULD YOU GO TO WAR FOR $1 TRILLION IN GOLD?

Consider the following war game: Two countries each have $1 trillion, and they are eyeing each other's gold. At the beginning of the game, the "strength" of each country's army is drawn randomly from a continuous uniform distribution and lies somewhere between 0 (very weak) and 1 (very strong). Each country knows its own strength but not that of its opponent. The countries observe their own strength and then simultaneously announce "peace" or "war."

If both announce "peace," then they each stay quietly in their own territory, with their own gold, so each "wins" $1 trillion.

If at least one announces "war," then they go to war, and the country with the stronger army wins the other's gold (i.e., the stronger country wins $2 trillion, and the other wins $0).

What is the optimal strategy of each country (declaring "peace" or "war") given its strength?

EXTRA CREDIT: What if the countries don't announce at the same time and instead one announces first and the other second? What if the value of winning the war were $5 trillion rather than $2 trillion?

Submitted by Juan Carrillo

SOLUTION:

The optimal strategy? Always declare war!

Why? Suppose the optimal strategy is to declare war when your strength is greater than some number X. This is a natural approach—the stronger your army, the more inclined you might be toward belligerence. If this is the optimal strategy, and if each player is facing an identical situation at the beginning of the game, then both players will play according to this threshold strategy. Let's assume they do. Here's where the game theory comes in.

If Country B declares war if and only if its strength is greater than X, Country A would do well to declare war whenever its strength is greater than $X/2$. How could it make sense for A to declare war if its army is only half as strong as B's army? Consider the two cases that follow from B's decision. When B's strength (unknown to A) is greater than X, it doesn't make any difference what A does because B will always declare war and A will be forced to fight. But in those situations where B's strength is less than X, A will of course win if its army is stronger than X, but there are many scenarios in which A's army will still be stronger than B's army even below X. Indeed, it profits by lowering its threshold to $X/2$. In half of those scenarios, it will be stronger than B, win the war, and double its gold cache. Country A would also still win in all those scenarios where it is especially strong, that is, when its strength is greater than X.

But if A is declaring war when its strength is greater than $X/2$, then B will do well to declare war when its strength is greater than $X/4$. But what's keeping A from dropping to $X/8$? It's bloodshed all the way down. They don't stop until they're both declaring war in every situation. Put another way, the only equilibrium "threshold" strength—X, in our notation—is 0. The strengths are always greater than 0, so both countries will always declare war.

The extra credit question asked what would happen if the declarations were made sequentially rather than simultaneously, and what would happen if winning the war was worth even more. It turns out neither makes any difference. Here's why: Suppose Country A must declare first. If it declares war, it doesn't matter what B does. But if it declares peace, it signals to B that its strength is low. So B will want to declare war in many of those situations. But if B did declare peace, A would have wanted to declare war, and so forth. So the optimal strategies collapse to always declaring war. And if the value of winning the war is increased from $2 trillion to $5 trillion, there are simply larger spoils at the end of the war, making declaring war even more attractive.

War. What is it good for? Nash equilibrium.

THE HOLIEST DAYS
ON THE PIIST CALENDAR

A certain religious calendar has a stretch of holiday Pi Days, during which adherents celebrate the number pi for several days and nights. You, however, have forgotten the starting and ending dates of Pi Days. You don't even know what month(s) the holiday will be in, but you do remember it is definitely not longer than two weeks. If the dates for Pi Days are presented in the format "MM/DD – MM/DD," and you treat that as a subtraction of fractions, you get the result 22/7 (about 3.14, or a decent approximation to the number pi). For example, January 1 through January 3 would be written "1/1 – 1/3," and if that expression is treated as a subtraction of fractions, the result is 2/3.

What are the starting and ending dates for Pi Days?

Submitted by Alex Jordan

SOLUTION:

Essentially, we are looking for a solution to $a/b - c/d = 22/7$, subject to certain restrictions on a, b, c, and d. We could ask our computer to do this for us, but where's the fun in that? Here's a pen and paper solution:

For starters, a and c are between 1 and 12 (they represent months) while b and d are between 1 and 31 (they represent days). Note also that $22/7$ is larger than 3, and a/b is larger than $22/7$, so $a/b > 3$. This implies that b cannot be greater than or equal to 4. We know b is 1, 2, or 3.

Because Pi Days last no longer than two weeks, we can now deduce that both dates are in the same month and, therefore, that $a = c$.

Because the right-hand side of our equation has a 7 in the denominator, one of the left-hand side's denominators must be divisible by 7. Since b is 1, 2, or 3, the second denominator, d, must be 7, 14, 21, or 28. But it can't be 21 or 28 because then Pi Days would be longer than two weeks. So d must be either 7 or 14.

At this point we know, thanks to a little rearranging, that $a(1/b - 1/d) = 22/7$, which is equivalent, after a little more multiplication, to $7a(d - b) = 22bd$. Noticing that the coefficient 22 on the right-hand side of the equation is divisible by 11, we can assume that 11 either divides a or it divides $(d - b)$. If 11 divides a, then a equals 11, and the equation reduces to $7(d - b) = 2bd$, with the restriction that b is 1, 2, or 3, and d is 7 or 14. In that case b can't be 1, because then the equation would be $7(d - 1) = 2d$, and $(d - 1)$ is relatively prime to d, meaning it doesn't share factors and is too big to divide 2. If b were 2, we must have $d = 14$ to make the left side even. But $11/2 - 11/14 = 22/7$ is false. Finally, b couldn't be 3, because the equation would be $7(d - 3) = 6d$, implying that 3 divides d, but the only options for d are 7 and 14. So 11 can't divide a. Therefore, it must be that 11 divides $(d - b)$.

Since Pi Days last at most two weeks, $(d - b)$ is 11. Since b is 1, 2, or 3, this rules out d being 7. So d must be 14, with b being 11 days earlier, at 3. Finally, solving for a in $a/3 - a/14 = 22/7$, we find $a = 12$.

Happy Pi Days—December 3 through December 14, we now know—to those who celebrate!

A SARTORIAL STUMPER

Three smart logicians are standing in a line, so that they can only see the logicians in front of them. A hat salesman comes along and shows the three logicians that he has three white hats and two black hats. He places one hat on each logician's head and hides the remaining hats.

He then says to the logicians, "Can anyone tell me what color hat is on her own head?" No one responds.

He repeats, "Can anyone tell me what color hat is on her own head?" Still no answer.

A third time: "Can anyone tell me what color hat is on her own head?" One of the logicians speaks up and gives the correct answer.

Who spoke, and what color hat is on her head?

Submitted by Milo Beckman

SOLUTION:

Name the logicians One, Two, and Three. Logician One is at the back of the line and can see the hats of Two and Three. Logician Two is in the middle and can only see the hat of Three. Logician Three can't see any hats.

There are eight arrangements of white and black hats we can consider. (Each logician could have a black hat or a white hat, and two times two times two is eight.) We can eliminate one of them right away: The salesman only has two black hats, so everyone knows that not everyone can be wearing a black hat. So we're left with seven possibilities for the hats of One, Two, and Three, respectively: WWW, WWB, WBW, BWW, WBB, BWB, and BBW.

When none of them knows the color of their hat when first asked, we can eliminate WBB. If One had seen two black hats in front of her, she'd have known that her own hat was white for sure. This leaves WWW, WWB, WBW, BWW, BWB, and BBW. Two and Three now know that they aren't both wearing black hats.

When none of them knows the color of her own hat when asked a second time, we can eliminate WWB and BWB. If Two sees Three wearing a black hat, Two knows that her hat must be white for sure. This leaves WWW, WBW, BWW, and BBW.

In all of these cases, Three's hat is white! So logician Three, despite not being able to see any hats, speaks up and declares—correctly—that her hat is white.

CAN YOU RIG THE
ELECTION . . . WITH MATH?

Imagine your job is to draw districts and you happen to be a member of the Dot Party. The grid below gives the locations of 25 voters in a region, which you must divide into five districts with five voters in each district. In each district, the party with more votes will win. The districts must be contiguous and not overlap (i.e., each square in a district must share an edge with at least one other square in the district). Can you draw the districts such that the Dot Party wins more districts than the Line Party?

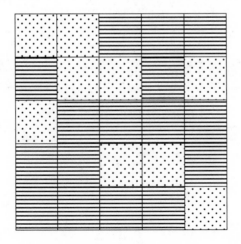

In the real world, of course, there aren't just 25 voters. Even if you can group neighborhoods together, the grid of voters in an entire state is going to be much larger, meaning that a computer program will probably be necessary to optimally gerrymander. Below is a rough approximation of Colorado's voter preferences, based on county-level results from the 2012 presidential election, in a 14-by-10 grid. Colorado has seven districts, so each would have 20 voters in this model. What are the most districts that the Line Party could win if you get to draw the districts with the same rules as above? What about the Dot Party? (Assume ties within a district are considered wins for the party of your choice.)

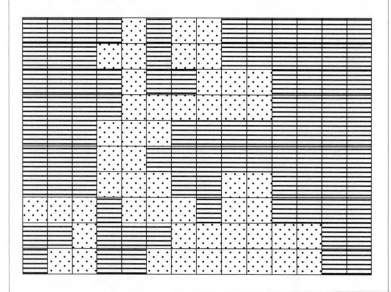

Submitted by Eli Ross

SOLUTION:

Even though Line Party voters outnumber Dot Party voters 16 to 9, you can draw five districts such that the Dot Party wins three and the Line Party only two. Democracy!

Although there are a few different ways to draw the districts, a common principle among them is "packing." The idea, if you're a Dot Party member, is to pack as many Line Party voters as possible into a couple of districts. The Line Party will win those handily but won't have enough voters in the remaining districts to win them. Here are two arrangements that will do the trick:

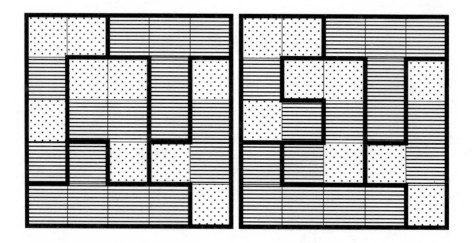

In both cases, the Line Party wins two districts by a landslide vote of 5 to 0, but the Dot Party wins the remaining three districts, each by a narrow vote of 3 to 2.

Let's turn to our rough approximation of the state of Colorado with seven districts. Specifically, you were tasked with finding the largest number of districts that either party could win if it had a sympathetic district-drawer. (In this case, Line voters outnumbered Dot voters 89 to 51.)

Given its hefty majority of voters in the state, the Line Party can win all seven of the districts. Here's a map that accomplishes that outcome:

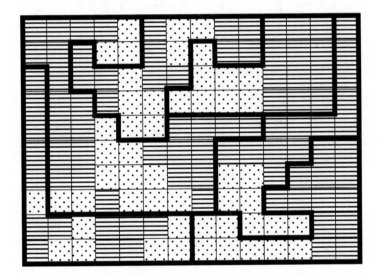

With some clever redistricting, the Dot Party can win at most five of the seven districts. (Each district has 20 voters, so it takes at least 10 voters to win one district. Dot has 51 voters, so five districts is its ceiling. We assumed that tied districts were won by the party of your choice.) Here's a map that achieves that outcome:

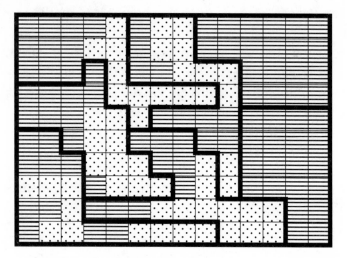

Gobs of Line voters are packed into a couple of districts, which Line wins handily, but its influence is diluted elsewhere. That allows Dot to eke out some narrow victories. Maps matter.

HERE'S $1 BILLION. CAN YOU WIN THE SPACE RACE?

You are the CEO of a space transport company in the year 2080, and your chief scientist tells you that one of your space probes has detected an alien artifact at the Jupiter Solar Lagrangian point.

You want to be the first to get to it! But you know that the story will leak soon and you only have a short time to make critical decisions. With standard technology available to anyone with liquid cash, a manned rocket can be quickly assembled and arrive at the artifact in 1,600 days. But with some nonstandard items you can reduce that time and beat the competition. Your accountants tell you that they can get you an immediate line of credit of $1 billion.

You can buy:

1. **Big Russian engines.** There are only three in the world and the Russians want $400 million for each of them. Buying one will reduce the trip time by 200 days. Buying two will allow you to split your payload and will save another 100 days.
2. **NASA ion engines.** There are only eight of these $140 million large-scale engines in the world. Each will consume 5,000 kilograms (kg) of xenon during the trip.

There are 30,000 kg of xenon available worldwide at a price of $2,000/kg, so 5,000 kg costs $10 million. Bottom line: For each $150 million fully fueled xenon engine you buy, you can take 50 days off of the trip.

3. **Light payloads.** For $50 million each, you can send one of four return-flight fuel tanks out ahead of the mission using existing technology. Each time you do so you lighten the main mission and reduce the arrival time by 25 days.

What's your best strategy to get there first?

Submitted by Robert Youngquist

SOLUTION:

If you buy one Russian engine ($400 million, saving 200 days), three ion engines ($450 million total, saving 150 days), and three light payloads ($150 million, saving 75 days), you will spend your $1 billion and get there 425 days earlier. This seems like the right answer—and it would be if the question were how to get there fastest.

But the challenge was to get there first. If you go with the option above, a competitor could also buy the same items and you would have no advantage. Here's one correct answer (there may exist variations on it that are also correct): Buy two Russian engines ($800 million, saving 300 days), one ion engine ($150 million, saving 50 days), and the rest of the world's xenon (25,000 kg for $50 million). Basically, you want to hoard the xenon. You will get there only 350 days earlier, but the best anyone else can do is to buy one Russian engine and four payloads to get there 300 days earlier. You are guaranteed to win by at least 50 days.

Xenon for days! Live long and prosper.

CAN YOU FIND THE
HONEST PRINCE?

You're the most eligible bachelorette in the kingdom, and you've decided to marry a prince. The king has invited you to his castle so that you may choose from among his three sons. The eldest prince is honest and always tells the truth. The youngest prince is dishonest and always lies. The middle prince is mischievous and tells the truth sometimes and tells lies the rest of the time. Because you will be forever married to one of the princes, you want to marry the eldest (truth-teller) or the youngest (liar) because at least you know where you stand with them. But there's a problem: You can't tell the princes apart just by looking, and the king will grant you only one yes-or-no question that you may address to only one of the brothers.

What yes-or-no question can you ask that will ensure that you do not marry the middle prince?

Submitted by Chris Horgan

SOLUTION:

Suppose you call the princes Al, Bob, and Chuck. Ask Al the following: "If I asked you if Bob was the middle brother, would you say yes?" If he says yes, marry Chuck and you're guaranteed to avoid the mischievous brother. If Al says no, though, marry Bob. (Poor Al.)

Why does asking a hypothetical question work better than asking the question outright? Let's break down all the possible true identities of Al, Bob, and Chuck, how they'd each answer the question, and what husband you'd choose to marry in all situations. There are six cases to consider.

AL	BOB	CHUCK	IS BOB ACTUALLY THE MIDDLE (MISCHIEVOUS) BROTHER?	AL'S RESPONSE TO THE HYPOTHETICAL QUESTION	WHO YOU PICK	YOUR RESULTING HUSBAND IS ..
Honest	Mischievous	Dishonest	Yes	Yes	Chuck	Dishonest
Honest	Dishonest	Mischievous	No	No	Bob	Dishonest
Mischievous	Honest	Dishonest	No	Yes or no	Bob or Chuck	Honest or dishonest
Mischievous	Dishonest	Honest	No	Yes or no	Chuck or Bob	Honest or dishonest
Dishonest	Honest	Mischievous	No	No	Bob	Honest
Dishonest	Mischievous	Honest	Yes	Yes	Chuck	Honest

In every case, you marry either the honest or dishonest prince, and you successfully avoid the mischievous one! That's because by asking a question about how the prince would answer a different, hypothetical question, you coerce the honest and dishonest princes into answering the same way, allowing you in both cases to avoid a life wedded to the mischief-maker.

Suppose Bob really is the middle brother. When you ask the honest brother how he would respond if you asked whether Bob was the middle brother, the honest brother will respond yes: He would say Bob is the middle brother, since Bob really is the middle brother. The dishonest brother would also say yes, because he always has to lie. What he's really saying is: "Yes, I'd tell you Bob is the middle brother, because when you actually asked me the question, I'd tell you Bob wasn't the middle brother, because I have to lie!" In other words, he's lying about whether he'd lie to you.

Another key is to ask a question of a person whom you will never marry, because the brother to whom you pose the question may well be the mischievous one.

THE STATE HIGHWAY ENGINEER AND HIS SUSPICIOUS, UNPAID INTERNS

Interstate 99 south of Riddler Springs consists of a straight, flat stretch of freeway with two lanes in each direction. There are no exits or entrances onto or off I-99 between Riddler Springs and Puzzlertown, the next city 20 miles to the south of Riddler Springs. There is, however, a road construction project going on in the southbound lanes beginning at a point 5 miles south of Riddler Springs. This project has reduced southbound travel to one lane for a few hundred yards, and the lane closure has produced traffic backups of a mile or more in the southbound lanes.

In order to learn more about the traffic jams, Ed, the state highway engineer, dispatched four of his unpaid interns, Andy, Barb, Chuck, and Di, to spend an hour alongside I-99 between 2 p.m. and 3 p.m. last Friday. Here are their reports:

ANDY: I stood 2 miles north of where the traffic jam begins with a radar gun and recorded the speed of every car going southbound. All the traffic at that point was moving along at 60 miles per hour, give or take 1 or 2 miles per hour.

BARB: I stood with Andy and counted the cars going southbound. For the whole hour, there was a steady flow of cars passing me going south at a rate of 80 cars per minute.

CHUCK: I stood 2 miles south of Barb and Andy, right at the point where the traffic backup began, a half-mile north of the point where the two southbound lanes merged into a single lane. The whole hour I was there, the backup in the southbound lanes stayed the same; it didn't get longer and it didn't get shorter.

DI: I stood along the road in the middle of the traffic backup, a quarter-mile south of Chuck and a quarter-mile north of the point where the two southbound lanes had to merge into a single lane. I had a radar gun, and the cars going south by me for the whole hour chugged along at a steady 4 miles per hour.

As soon as Ed read these four reports, he knew at least one of them had to be incorrect. How did he know? Whom should he fire?

Submitted by Dave Moran

SOLUTION:

If Chuck is correct that the traffic backup wasn't getting longer or shorter, then the frequency of cars going by Andy and Barb (which Barb said was 80 cars per minute) must be the same as the frequency of cars going by Di, who is in the middle of the backup, because there are no places for cars to exit or enter between Barb and Di. (This is analogous to an observer of electromagnetic radiation who is in a dielectric medium, such as underwater: The observer above the water will see the same frequency of wave crests as the observer who is underwater, even though the light moves more slowly in the water because the wavelength—like the distance between cars—will shorten in the dielectric medium.)

If there are 80 cars per minute going by Di in two lanes, each lane must have 40 cars passing by per minute. That means a car must pass by in each lane every 1.5 seconds. To put it another way, if cars are literally bumper-to-bumper, the length of one car must pass by in 1.5 seconds in order for 40 cars to pass per minute in each lane. But Di reported the cars were going by her at 4 miles per hour, which translates to just under 6 feet per second, so a car going 4 miles per hour will travel less than 9 feet in 1.5 seconds. Because 9 feet is less than the length of even a compact car, it is not possible for 40 cars to pass Di in each lane per minute at a speed of 4 miles per hour.

Note that it's not possible to say which or how many of the four reports are wrong. If Barb's report as to the traffic frequency is too high, then the other reports could all be correct. If Chuck is wrong and the traffic backup is getting longer, then Barb's report as to frequency and Di's report as to the speed of the cars in the jam could both be correct because there would be fewer cars going by Di per minute than are going by Barb. And if Di's report as to the speed of the cars in the backup is too low, then the other reports could all be correct. The answer, then, is that Ed should fire Andy, since his report is actually irrelevant to the problem; it doesn't matter how fast the cars are going before the jam begins.

WHAT'S THE BEST
WAY TO DROP
YOUR SMARTPHONE?

You work for a tech firm developing the newest smartphone that supposedly can survive falls from great heights. Your firm wants to advertise the maximum height from which the phone can be dropped without breaking.

You are given two smartphones and access to a 100-story tower from which you can drop either phone from whichever story you want. If it doesn't break when it falls, you can retrieve it and use it for future drops. But if it breaks, you don't get a replacement phone.

Using the two phones, what is the minimum number of drops you need to ensure that you can determine precisely the highest story from which a dropped phone does not break? (Assume you know that it breaks when dropped from the very top.) What if, instead, the tower is 1,000 stories high?

Submitted by Laura Feiveson

SOLUTION:

For 100 stories, the smallest number of drops needed is 14. Here's a strategy that does it: Name your phones A and B. Drop phone A from floors 14, 27, 39, 50, 60, 69, 77, 84, 90, 95, and 99 for as long as it doesn't break. When it does break, test the floors between the previous drop (where it didn't break) and the current floor (where it just broke). To do this, go up one floor at a time with phone B until either it breaks or you reach the floor just below the one where phone A broke. For example, suppose that phone A broke when dropped from floor 39 (after having survived the two previous falls from floors 14 and 27). You would then take phone B and drop it from floor 28 and then move up one floor at a time until it broke. Suppose that you get to floor 38 without phone B breaking—then you know that floor 38 is the highest floor from which you can drop a phone without it breaking!

An important feature of this strategy is that the most drops you will need is 14 regardless of where phone A breaks. If phone A breaks with your first drop at floor 14, then you need at most an additional 13 drops from phone B to determine the "breaking floor"—that adds up to a maximum of 14 drops, one from phone A and 13 from phone B. But if instead phone A doesn't break until floor 90, you will need only a maximum of five drops from phone B—again, a maximum of 14 drops: nine from phone A and five from phone B. Intriguingly, the maximum number of drops you need is the same regardless of where phone A breaks. You can't "improve" upon the strategy.

To generalize this strategy to more floors, note that the strategy follows a precise pattern in the sequence of floors from which you should drop phone A: Start with 14, then add 13, then add 12, then add 11, and so on. The solution for 100 floors was 14 because 14 is the smallest number such that $1 + 2 + 3 + \ldots + n \geq 100$. This classic sequence problem has a well-established solution: $1 + 2 + 3 + \ldots + n = n(n + 1)/2$. To solve for 1,000 floors, or any number of floors f, just find the smallest n such that $n(n + 1)/2 \geq f$. For the case of 1,000 floors, the solution is 45 drops. Note that the exponential nature of the series means that even though 1,000 is

10 times 100, you need only roughly three times the number of drops to find the "breaking floor." For 10,000 floors, you would need to have only 141 drops.

Finally, one beautiful extension is to think about how many total drops you would need if you had three smartphones rather than two. In fact, it is possible to generalize even more and to come up with a relatively simple formula that determines the number of drops you would need for m phones and f floors. This is left as an exercise for the reader.

WILL THE DWARVES SURVIVE?

A giant troll captures 10 dwarves and locks them up in his cave. That night, he tells them that in the morning he will decide their fate according to the following rules:

1. The 10 dwarves will be lined up from shortest to tallest so each dwarf can see all the shorter dwarves in front of him but cannot see the taller dwarves behind him.
2. A white or black dot will be randomly put on top of each dwarf's head so that no dwarf can see his own dot but they can all see the tops of the heads of all the shorter dwarves.
3. Starting with the tallest, each dwarf will be asked the color of his dot.
4. If the dwarf answers incorrectly, the troll will kill the dwarf.
5. If the dwarf answers correctly, he will be magically, instantly transported to his home far away.
6. Each dwarf present can hear the previous answers but cannot hear whether a dwarf is killed or magically freed.

The dwarves have the night to plan how best to answer. What strategy should be used so that the fewest dwarves will die, and what is the maximum number of dwarves that can be saved with this strategy?

Submitted by Corey Fisher

SOLUTION:

Nine of the 10 dwarves can be saved for sure and, with a little luck, all 10 will escape the troll's clutches. How? The dwarves agree on the following plan: The first dwarf who is the tallest will risk life and limb to save the others. Since he has no information to go on to determine his own dot's color, he can use his guess to inform the others. The dwarves agree that if the number of white dots the tallest dwarf sees is even, he should say "white," and if it's odd, he should say "black."

That first dwarf only has a 50–50 chance of survival, but all of his compatriots will now survive for sure because they know why he said the color he said. Suppose the first dwarf says "white," meaning he sees an even number of white dots. Then it's the second dwarf's turn. If he also sees an even number of white dots, then he knows for sure that his dot is black. If, instead, he sees an odd number of white dots, then he knows for sure that his dot is white. Based on the responses of the first two dwarves, the third can then also determine the "evenness" or "oddness" of the remaining white dots. If what he sees matches that, his dot must be black; if not, white, and so on.

Regardless of how many dwarves there are (say there are N), at least $N-1$ can be saved for sure, and all can be saved half the time! Your fellow dwarves thank you, tallest dwarf.

HOW BADLY CAN
A CAR SALESMAN
SWINDLE YOU?

You want to buy a specific car, whose fair price is N. You have absolutely no idea what N is and the dealer, sadist that he is, won't tell you. The dealer enjoys a good chase, though, so without directly revealing the value of N, he takes five index cards and writes down a number on each card: N, $N + 1{,}000$, $N + 2{,}000$, $N + 3{,}000$, and $N + 4{,}000$. Important: The guy is sadistic but he's not a math major. The numbers on the cards are numbers, not algebra equations. (In other words, if the car costs \$20,000, he has five cards labeled \$20,000, \$21,000, etc.)

He presents the cards to you, one at a time, in random order. (For example, if the price on the second card is \$1,000 more than the first, you can't be sure if those are the two smallest prices, the two largest, or somewhere in between.) Each time he shows you a card, you must either pay the price on that card or ask to see the next card. You cannot go back to previous cards. If you make it to the fifth and final card, then that number is what you must pay.

If you know the parameters of the dealer's game and you play it optimally, how much should you expect to pay on average above the fair price N?

Submitted by Zach Wissner-Gross

SOLUTION:

If you play the car dealer's game optimally, you can expect, on average, to pay $900 above the fair price.

To explain why, let's step back for a second. If you adopted a naive strategy of simply accepting the first card, say, you'd expect to pay $2,000 over the fair price—simply the average of all the possible markups. That's our baseline, but we can do significantly better. The reason we can do better is that we can learn new information by asking to see additional cards. Put another way: Never pick the first card.

Call the cards—which, remember, display the numbers N, $N + 1,000$, $N + 2,000$, $N + 3,000$, and $N + 4,000$—A, B, C, D, and E, respectively.

The optimal approach depends on the interval between the numbers on the first two cards. There are four possible intervals. Here is one strategy that achieves the smallest expected overpayment:

1. **The first two cards are $4,000 apart.** You now know for sure that you saw A and E. If you saw A then E (which occurs 1 out of 20 times), then you can wait for B and pay $N + 1,000$. If you saw E then A (which also occurs with probability 1/20), then you obviously take A, the lowest card, and pay the fair price N. The average price in this scenario is $N + 500$.

2. **The first two cards are $3,000 apart.** These may be A and D or B and E. The best strategy in this scenario is to wait for the card that is $1,000 higher than the lower of the first two or better. There are two subcategories to this interval.

   a. If you saw A then D, you are guaranteed, at some point, to get B, which is $1,000 higher than A. If you saw B then E, you are going to see either A or C with one-half probability (because you'll pick whichever you get first). Averaging these results, you should expect to pay $N + 1,000$. One out of 10 times the third card will deliver this precise result.

b. If you saw D then A or E then B, you should choose the second card to pay $N + 500$ on average (also 1 out of 10 times).

3. The cards are \$2,000 apart. The best strategy for the three remaining cards (they can be BDE, ACE, or ABD) is characterized by the following:

a. If you see a card that is lower than the first two, take it.

b. If your first four cards are in consecutive order, and the fourth one is the second lowest, take it (e.g., if you saw $CADB$ or $DBEC$, take the fourth card).

c. Otherwise, wait until the end.

This strategy results in you paying $N + 1{,}000 \times (8/9)$ on average. This is better than taking the second card even if the second card is \$200 lower than the first (the expected amount would be $N + 1{,}000$ if you take it). This case occurs 3 out of 10 times.

4. The cards are \$1,000 apart. The strategy for the remaining three cards is the following:

a. If you see a card lower than the first two, take it.

b. If you see both A and E, wait until the lowest available.

c. Otherwise, wait until the end.

This strategy results in the expected payment of $N + 1{,}000 \times (13/12)$. This is again better than taking the second card. This case occurs 4 out of 10 times.

The final tally for the overall expected payment is

$$\left(\frac{1}{20}\right)(0) + \left(\frac{1}{20}\right)(1{,}000) + \left(\frac{1}{10}\right)(1{,}000) + \left(\frac{1}{10}\right)(500)$$

$$+ \left(\frac{3}{10}\right)\left(\frac{8}{9}\right)(1{,}000) + \left(\frac{4}{10}\right)\left(\frac{13}{12}\right)(1{,}000) = 900$$

You pay, on average, \$900 more than the fair price, which is not bad at all! (Actually, it is terrible, but you did the best you could and it's definitely better than the worst possible price.)

ROCK, PAPER, SCISSORS, DOUBLE SCISSORS

Who doesn't love Rock, Paper, Scissors? But what if we can make it even better? Besides the usual three options that players have—rock, paper, or scissors—let's add a fourth option, double scissors, which is played by making a scissors with two fingers on each side (like a Vulcan salute). Double scissors, being larger and tougher, defeat regular scissors, and just like regular scissors they cut paper and are smashed by rock. The three traditional options interact just as they do in the standard game. A match of rock, paper, scissors, double scissors is always played best two out of three (or, more precisely, first to win two throws, since there can be an unlimited number of ties).

To keep things interesting, there is one wildcard rule: If your opponent throws paper and you throw regular scissors, you immediately win the match regardless of the score. What is the optimal strategy at each possible score (0–0, 1–0, 0–1, and 1–1)? (You can ignore any ties.) What is the probability of winning the match given a 1–0 lead?

Submitted by Patrick Coate

SOLUTION:

In a traditional game of Rock, Paper, Scissors, the only optimal strategy is to randomize equally between the three objects—to play each with probability 1/3. Any other strategy can be exploited by a savvy opponent. If you favored scissors as your go-to throw, for example, your opponent could always throw rock and would best you in the long run.

A similar argument applies to our modified game. But, first, notice that it is never optimal for a player with one win to play regular scissors. With one win already in the bag, there is no benefit from the exceptional rule where a win with regular scissors wins the match automatically, and double scissors beats everything that scissors does, plus scissors itself. In game theory parlance, double scissors *dominates* regular scissors. Therefore, if the score is tied 1–1, our modified game is reduced to rock, paper, double scissors, and the optimal strategy is to play each with probability 1/3. That's one piece of our solution done!

What about a match with a score of 1–0? Suppose you are in the lead and your opponent is trailing. You both know that you'll select from rock, paper, and double scissors (because regular scissors is dominated) and your opponent will select from rock, paper, scissors, and double scissors. Let's make a table showing all your possible plays and your probability of winning given those plays. Call your chances of winning with a 1–0 lead X, which is what we're searching for to solve this problem.

SOLVING FOR YOUR NEXT THROW WHEN YOU'RE UP 1–0

	Your chance of winning the match if you throw . . .		
OPPONENT'S THROW	ROCK	PAPER	DOUBLE SCISSORS
Rock	X	1	.5
Paper	.5	X	1
Scissors	1	0	1
Double scissors	1	.5	X

For instance, if you play rock and your opponent plays double scissors, you smash your opponent's double scissors, take a 2–0 lead, and win the match—so your probability of winning the match is 1. If you play rock and your opponent plays paper, your opponent covers your rock and ties the match 1–1. Because you're both playing optimally, a tied match must give you both a 50–50 chance of winning it all. If you play paper and your opponent plays scissors, your opponent cuts your paper and wins the match automatically, thanks to the special rule—so your probability of winning the match is 0.

But there are some Xs in the table that we don't yet know! Those are the chances that you win when you play optimally with your 1–0 lead. Let's say, in your optimal strategy, that you play rock with some probability R_1, double scissors with probability DS_1, and paper with probability P_1. No matter what your opponent plays, playing according to this strategy should deliver you your optimal 1–0 win chances of X. So, for example, if your opponent plays rock, we get an equation like this from the table above:

$$(R_1)(X) + (DS_1)(.5) + (P_1)(1) = X$$

If your opponent throws rock: In those times you wind up throwing rock, you win with probability X; in those times you throw double scissors, you win with probability .5; and in those times you throw paper, you win with probability 1. The same holds true for the other three things your opponent might throw, which gives us a system of equations:

$$(R_1)(1) + (DS_1)(X) + (P_1)(.5) = X$$
$$(R_1)(1) + (DS_1)(1) + (P_1)(0) = X$$
$$(R_1)(.5) + (DS_1)(1) + (P_1)(X) = X$$

Your opponent is trying to solve a very similar problem, where your opponent is playing rock with probability R_2, double scissors with probability DS_2, scissors with probability S_2, and paper with probability P_2. So if you play rock, for example, your opponent knows that

$$(R_2)(X) + (DS_2)(1) + (P_2)(.5) = X$$

The solution to this big system of equations gives $X \approx .73$. With a win in hand, you're likely to win the match nearly three-quarters of the time. The equations also suggest that you should play rock, double scissors, and paper with probabilities .40, .33, and .27, respectively. Your opponent should play rock, double scissors, scissors, and paper with probabilities .55, 0, .21, and .25, respectively. Three more pieces of our solution done! All that's left is to figure out the strategy at the beginning of the game—score 0–0. We can make another table to help with solving the problem.

SOLVING FOR YOUR NEXT THROW WHEN YOU'RE TIED 0–0

	Your chance of winning the match if you throw . . .			
OPPONENT'S THROW	ROCK	PAPER	SCISSORS	DOUBLE SCISSORS
Rock	.5	X	$1 - X$	$1 - X$
Paper	$1 - X$	.5	1	X
Scissors	X	0	.5	X
Double scissors	X	$1 - X$	$1 - X$	.5

The solution process is the same as above, too, but I'll spare you all the algebra. The optimal strategy that pops out of the math is that at score 0–0, you should both play rock, double scissors, scissors, and paper with probabilities .52, 0, .24, and .24, respectively. One interesting finding: No one should ever play double scissors before a win or regular scissors after a win.

It's like the old saying goes: In the world of double scissors, the rock is king.

HOW LONG WILL
CHAOS REIGN IN
THIS GAME OF TAG?

On a sunny summer day, you've gone to the park with your friends. After the picnic is eaten, you decide to enjoy the weather by playing a nice game of chaos tag. For those uninitiated in the joys of chaos tag, you explain the following rules:

1. Any group of two or more people can play. All players are active at the start of the game.
2. Active players can run around and tag other active players.
3. A player who is tagged becomes inactive and must sit on the spot where he or she was tagged.
4. An inactive player becomes active again when the player who tagged him or her is tagged.
5. Victory is achieved by being the only remaining active player.

Suppose N people are at the park that day. If all active players are equally likely to tag someone and any of the possible targets are equally likely to be tagged, how long will the game last on average, as measured in tags?

Submitted by Ryan Tavenner via Austin Shapiro

SOLUTION:

The biggest problem with chaos tag is that as the number of players increases, the number of tags the game will require increases very fast. In a game with three players, the expected solution is three tags. Why? Say the players are Alyssa, Barry, and Carl. Let's say that the game begins with Alyssa tagging Carl. Half the time after that, Alyssa tags Barry, too, and the game ends in two tags with Alyssa victorious. The other half of the time, though, Barry tags Alyssa, Carl gets new life, and the game is back to where it began. Call the expected number of tags in this game E. The game always begins with one tag, ends on the next tag half the time, and returns to where it began half the time, like so:

$$E = 1 + \left(\frac{1}{2}\right)(1) + \left(\frac{1}{2}\right)(E)$$

Solving for E quickly gives $E = 3$. Continuing that logic, you'll find that a game with five players takes 15 tags on average. A game with 10 players takes 511 tags. A tag every 10 seconds and you're looking at game lasting an hour and a half.

Imagine, however, a game with 100 players. In the extremely unlikely scenario of two very skilled players each tagging 49 people, the "final" tag would immediately put 49 players back into play. If the best players are only good enough to tag five people before being tagged, the tag totals begin to skyrocket. A 100-player game is expected to take 633,825,300, 114,114,700,748,351,602,687 tags. This game would take trillions of times longer than the age of the universe!

But here is a nifty analytical approach to solving the problem that fits on a single page: Each player P is responsible for some number $n(P)$ of currently inactive players. For example, at the start of the game $n(P) = 0$ for all players since no one has been tagged, and at the end of the game $n(Winner) = N - 1$. Define the "score" of player P to be $2^{n(P)} - 1$, and define the "total score" of the current game to be the sum of the scores of the players. So the total score at the beginning of the game is 0 and the total score at the end is $2^{N-1} - 1$. Suppose we are in the midst of the game and we consider what is going to happen next. In this scenario, P and Q are two active players. If P tags Q, the change in the score is $2^{n(P)} - 2^{n(Q)} + 1$, while if Q tags P, the change in the score is $2^{n(Q)} - 2^{n(P)} + 1$. Note that the average of these two values is 1. Since these two options are equally likely, according to the assumptions of the problem, and in fact all options occur in pairs like this, we see that the expected value of the score goes up by 1 each time. It follows that the expected remaining duration of the game starting from any given situation S will be Score(EndState) – Score(S). Thus, the expected length of the game will be Score(EndState) – Score(BeginState) = $2^{N-1} - 1$.

The average numbers of tags for each number of players are all very close to powers of two! That is, they almost, but not quite, match a simple doubling pattern of 1, 2, 4, 8, 16, 32, 64, and so on. From that observation, even if we had no idea where to start algebraically, we could posit our general solution for N players: $2^{N-1} - 1$ tags.

WHAT'S UP WITH YOUR MISANTHROPIC NEIGHBORS?

The misanthropes are coming! Suppose there is a row of houses, N houses long, in a new, initially empty development.

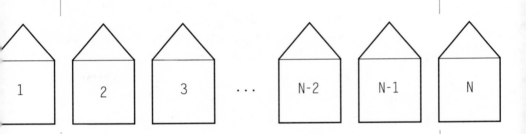

Misanthropes are moving into the development one at a time. Because misanthropes dislike other people, they each select an empty house at random where nobody lives next door. They keep arriving until no acceptable houses remain.

What's the expected fraction of occupied houses as the development gets larger, that is, as N goes to infinity?

Submitted by Jim Ferry

SOLUTION:

Since a misanthrope will only move into an empty house with no neighbors, the most tightly packed neighborhood, asymptotically, will have someone in one out of every two houses, while the emptiest neighborhood will have someone in one out of three houses. This suggests the Richard Feynman ploy of immediately declaring the answer to be $1/e$. Feynman, a Nobel Prize–winning theoretical physicist, was known to wow mathematical colleagues by instantly solving complicated computations, and there is a famous story about him calculating numerous powers of the number e. But is this correct? No.

The correct answer in this case is $(1 - e^{-2})/2$, or approximately $.432$. About 43.2 percent of the houses are expected to be occupied. Sorry, Feynman.

Why? Deep breath . . . It's math time!

Let a_n be the expected number of misanthropes in a row of n houses—this is the number we're looking for to solve the puzzle. Clearly $a_0 = 0$ and $a_1 = 1$. If there are no houses, there will be no people, and if there is one house, it will be occupied. But it quickly gets more complicated.

For larger n, real neighborhoods with more than one house, let's go misanthrope by misanthrope. The first misanthrope picks house k (simply one of the houses from 1 to n) with probability $1/n$, yielding an expected value a_n of 1 (due to that first misanthrope) plus contributions due to the misanthropes who move into the rows of houses to its left and right: houses 1 through $k - 2$ and $k + 2$ through n. (Remember that no one will now move into house $k - 1$ or $k + 1$.) We can now write an equation called a recursion—a relationship between the expected number of misanthropes in different segments of the neighborhood as the neighborhood grows. Specifically, the expected number of misanthropes in the neighborhood is one plus the expected number of misanthropes in all the other livable houses, or $a_n = 1 + (2/n) \sum_{j=1}^{n-2} a_j$.

Letting $s_n = \sum_{j=1}^{n} a_j$, we can rewrite this recursion as

$$a_n = 1 + (2/n)\, s_{n-2}$$

and

$$s_n = s_{n-1} + a_n$$

The good news is we don't really want to solve this system of equations—we just want to find the limit of a_n/n as n goes to infinity—that is, the proportion of the houses that will be filled as the neighborhood gets bigger and bigger. Therefore, we introduce generating functions $A(x) = \sum_{n=0}^{\infty} a_n x^n$ and $S(x) = \sum_{n=0}^{\infty} s_n x^n$. Generating functions are simply used to describe infinite series of numbers, such as our infinite series of growing neighborhoods. We can then multiply each side of the above recurrences by x^n and sum from $n = 0$ to infinity to obtain

$$A(x) = 1/(1 - x) + 2F(x)$$

where $F'(x) = xS(x)$ and

$$S(x) = xS(x) + A(x)$$

Therefore, $A(x) = (1 - x)S(x)$, and differentiating the first equation yields

$$(1 - x)S'(x) - S(x) = 1/(1 - x)^2 + 2xS(x)$$

with the initial condition $S(0) = 0$ (which enforces $s_0 = 0$). The solution to this equation is $S(x) = (1 - e^{-2x})/(1 - x)^3$, so

$$A(x) = (1 - e^{-2x})/(1 - x)^2$$

Therefore the asymptotic growth rate of a_n is $a_n \sim ((1 - e^{-2})/2)n$. That is, the limit of a_n/n as $n \to \infty$ is, indeed, $(1 - e^{-2})/2$. In expectation, about 43.2 percent of the houses will be occupied—closer to the tightest possible neighborhood than the emptiest.

YOU'RE THE NEXT CONTESTANT ON GUESS YOUR HAT!

You and six friends are on a hit game show that works as follows: Each of you is randomly given a hat to wear that is either black or white. Each of you can see the colors of the hats that your friends are wearing but cannot see your own hat. Each of you has a decision to make. You can either attempt to guess your own hat color or pass. If at least one of you guesses correctly and none of you guess incorrectly, then you win a fabulous, all-expenses-paid trip to the International Hat Expo. If anyone guesses incorrectly or everyone passes, you all lose. No communication is possible during the game—you make your guesses or passes in separate soundproof rooms—but you are allowed to confer beforehand to develop a strategy.

What is your best strategy? What are your chances of winning?

Submitted by Jared Bronski

SOLUTION:

It is very tempting to guess that your chances are 50–50—and many readers did when we first posed this problem. Someone has to guess right, she has to guess the color of a hat she cannot see, and that hat is randomly black half the time and white half the time, after all. But, remarkably, you can do a lot better as a team.

Your best chances of winning are 7/8, or 87.5 percent. You can achieve that with the optimal strategy described below.

To make the problem a bit more intuitive, first consider a simpler example where it's just you and two friends on the game show. You can win this version 75 percent of the time. The simple strategy that does it is this: Each player looks at the hats of the other two. If she sees one hat of each color, she passes. If she sees two of the same color (e.g., white), she guesses that hers is the other color (e.g., black). Essentially you are betting that the hats are not all the same color. At least one player will always see two hats of the same color, so at least one player always guesses. If the hats are WWB or BBW, then exactly one player guesses correctly. If the hats are all the same color, BBB or WWW, then everyone guesses incorrectly— these arrangements occur 25 percent of the time. This strategy "stacks" the incorrect guesses and "spreads out" the correct ones: Either one person guesses correctly or three people guess incorrectly.

Now back to our main problem, where seven of you are on the show. Things get much more complicated, but importantly the solution has the exact same flavor—either one person guesses correctly or all seven people guess incorrectly.

To keep track of everyone, let's assign each player a number. It turns out to be easiest if we do this in binary numbers, as we'll see shortly. So the seven of you get numbered like so:

Anna 001

Ben 010

Clarice 011

Doug 100

Edna 101

Fred 110

Georgina 111

Each player looks around and takes note of all the players wearing black hats. One player then carries out a small programmatic algorithm by "XORing" the numbers of all the black hat wearers to get a final count. (XOR is a logical operation, short for "exclusive or," which returns true if exactly one of its inputs is true. In binary, it's bitwise addition without carrying: "1 XOR 0" is 1, "1 XOR 1" is 0, and so on.) If the final count is 0, the player guesses black. If the count is his or her own number, the player guesses white. If the count is anything else, the player passes.

For example, suppose Anna, Ben, and Georgina have black hats and everyone else has a white hat. Anna sees two black hats—Ben's and Georgina's—so her count is 101 (010 XOR 111), and she passes. Similarly, Ben's count is 110 (001 XOR 111) and he passes, and Georgina's count is 011 (001 XOR 010) and she passes. The rest of the players—Clarice, Doug, Edna, and Fred—all get a count of 100. (To perform the XOR operation with three numbers just go digit by digit again. For example, 001, 010, and 011 XOR to 100, and the rightmost digits XOR to 0 because we don't carry. Ditto the middle digit. And the leftmost digits add to 1.) This is Doug's number, so three of them pass while Doug guesses white. Doug is correct!

When all's said and done, this approach gives a result quite similar to the simple three-player example: Either one person guesses correctly or all seven people guess incorrectly. Essentially you are betting that if you XOR together the numbers of the black hat wearers, the answer is not 0. The answer can be anything from 000 to 111, and all are equally likely, so this is a one-in-eight chance. Your team wins seven times out of eight.

THE PUZZLE OF
THE LONESOME KING

The childless King of Solitaria lives alone in his castle. The lonesome king one day offers one lucky subject the chance to be prince or princess for a day. The loyal subjects leap at the opportunity, having heard tales of the king's opulent castle and decadent meals. The subjects assemble on the village green, hoping to be chosen.

The winner is chosen through the following game. In the first round, every subject simultaneously chooses a random other subject on the green. (Of course, it's possible that some subjects will be chosen by more than one other subject.) Everybody chosen is eliminated. (Not killed or anything, just sent back to their hovels.) In each successive round, the subjects who are still in contention simultaneously choose a random remaining subject, and again everybody chosen is eliminated. If there is eventually exactly one subject remaining at the end of a round, he or she wins and heads straight to the castle for fêting. However, it's also possible that everybody could be eliminated in the last round, in which case nobody wins and the king remains alone. If the kingdom has a population of 56,000 (not including the king), is it more likely that a prince (or princess) will be crowned or that nobody will win?

Submitted by Charles Steinhardt

SOLUTION:

If there are 56,000 subjects, the chances that exactly one remains and moves into the castle are a touch less than one-half, or about 48 percent.

The answer to this problem, for kingdoms of various populations, exhibits some very strange behavior. The chance that someone wins (rather than a bunch of subjects all being simultaneously eliminated in the final round) hovers around 50 percent but never converges exactly to it. Rather, the probability oscillates around one-half, like a sine wave, sometimes slightly higher, sometimes slightly lower, depending on the exact number of subjects on the green. There does not seem to be an intuitive explanation for this weird behavior—it's a bit of a mystery! Even professional mathematicians are still at work on this problem. In the academic literature, a situation like this is sometimes known, darkly, as "group Russian roulette."*

I confess that this "solution" is not terribly satisfying. However, it is nourishing to find such beauty and complexity hidden just beneath the surface of such a simple situation.

---

* See, for example, "The Asymptotics of Group Russian Roulette" by Tim Van De Brug, Wouter Kager, and Ronald Meester—a dense, 26-page paper proving that the answer to this question oscillates around 50 percent, never settling.

PROBABILITY

The enchanting charms of this sub-
lime science reveal themselves in all
their beauty only to those who have
the courage to go deeply into it.

—CARL FRIEDRICH GAUSS

WILL YOU BUST
THE GHOSTS OR
DESTROY THE WORLD?

Twenty ghostbusters are on their annual camping retreat. Two of them, Abe and Betty, have discovered that another pair, Candace and Dan, are in fact ghosts posing as ghostbusters. Abe and Betty hatch a plan: When all 20 campers are sitting in a circle around the campfire, Abe will fire his proton pack at Candace, and Betty will simultaneously fire her proton pack at Dan, annihilating the ghosts. However, if two proton streams cross, it means the end of all life on Earth.

If the ghostbusters are arranged randomly around the fire, what are the chances that Abe and Betty will cross streams?

Submitted by Max Weinreich

SOLUTION:

The chances are 1/3.

There are 20 ghostbusters, but we only really care about four of them—Abe, Betty, Candace, and Dan. The position of the other 16 won't affect the possible crossing of the streams, so let's ignore them. (Sorry, you 16 irrelevant ghostbusters.)

Fix Abe's spot at the campfire—say he's on the north side. There are then three places his co-ghostbuster Betty can sit—east, west, or south. The ghosts, Candace and Dan, will sit in the other two seats. There are $3 \times 2 \times 1$ or 6 possibilities for seating in the east, west, and south seats. In exactly two of these arrangements—the two in which Candace occupies the south seat, forcing Abe to fire across the circle—the ghostbusters will cross their proton streams. Each of these six arrangements is equally likely, so the chances of stream-crossing disaster are $2/6 = 1/3$.

CAN YOU SURVIVE THIS DEADLY BOARD GAME?

While traveling in the Kingdom of Arbitraria, you are accused of a heinous crime. Arbitraria decides guilt or innocence not through a court system but through a board game. It's played on a simple board: a track with sequential spaces numbered from 0 to 1,000. The 0 space is marked "start," and your token is placed on it. You are handed a fair six-sided die and three coins. You are allowed to place the coins on three different (nonzero) spaces. Once placed, the coins may not be moved.

After placing the three coins, you roll the die and move your token forward the appropriate number of spaces. If, after moving the token, it lands on a space with a coin on it, you are freed. If not, you roll again and continue moving forward. If your token passes all three coins without landing on one, you are executed. On which three spaces should you place the coins to maximize your chances of survival?

EXTRA CREDIT: Suppose there's an additional rule that you cannot place the coins on adjacent spaces. What is the ideal placement now? What about the worst spaces—where should you place your coins if you're making a play for martyrdom?

Submitted by James Kushner

SOLUTION:

To give yourself the best shot at staying alive, place your coins on spaces 4, 5, and 6. You'll survive about 79.4 percent of the time.

A little intuition here goes a long way. For starters, you'll definitely want to place a coin on space 6. It's the best space because it maximizes the number of rolls that could land on it. You could roll a 6 and hit it in one shot, for example. If you don't get there in your first roll, you're guaranteed to have at least one more shot at landing there. Space 5 is another great candidate for the same reason. You could hit it on your first roll, of course, but if you roll something less than 5, you'll get another chance to save your life on that space.

You can confirm this intuition with math, of course. Once you've got your coins on spaces 5 and 6, you struggle with whether to place your final coin on space 4 or on space 7. Here's how you can decide.

For 4-5-6, here are the probabilities you win if you . . .

> Roll a 4, 5, or 6: .5
> Roll a 3: 1/6*(.5) = .083
> Roll a 2: 1/6*(.5 + .083) = .097
> Roll a 1: 1/6*(.5 + .083 + .097) = .113

for a total probability of about .794.*

For 5-6-7, here are the probabilities you win if you . . .

> Roll a 5 or 6: .333
> Roll a 4: 1/6*(.5) = .083
> Roll a 3: 1/6*(.5+ .083) = .097
> Roll a 2: 1/6*(.5 + 1/6*1/2 + 1/6*.097) = .113
> Roll a 1: 1/6*(.794) = .132

for a total probability of about .758. So 4-5-6 is little bit better.

---

* The numbers in the parentheses are the probabilities that your next roll lands you on one of your tokens.

The length of the board is a bit of a red herring. The answer would be the same if the board were 100 or 1,000,000 spaces long. The odds of landing on any given space are roughly equal—about 2/7—and it wouldn't much matter where you placed your coins.

If you're making a play for martyrdom, placing the coins on spaces 1, 2, and 7 minimizes your chances of survival. You'll live just 47.5 percent of the time. If adjacent coins aren't allowed, spaces 6, 8, and 10 are your best bets to survive, and spaces 1, 3, and 7 are best if you have a death wish.

WILL YOU (YES, YOU) DECIDE THE ELECTION?

You are the only sane voter in a state with two candidates running for the Senate. There are N other people in the state, and each one votes completely randomly! All voters act independently and have a 50–50 chance of voting for either candidate. What are the odds that your vote changes the outcome of the election in favor of your preferred candidate? More importantly, how do these odds scale with the number of people in the state? For example, if twice as many people lived in the state, how much would your chances of swinging the election change?

Submitted by Andrew Spann

SOLUTION:

If the N voters who aren't you vote randomly and independently for one of two candidates, who each have a 50 percent chance of winning, and you vote for your preferred candidate, there is about a $\sqrt{\frac{2}{N\pi}}$ chance that your vote will be decisive.

Why? Your vote will be decisive if and only if half of the N voters vote for one candidate and half of them vote for the other candidate. (For simplicity, let's assume N is even. If N is odd, the best your vote can do is to move the election into a tie. You can tweak the formula above slightly to reflect that possibility, but the distinction won't make any meaningful difference when N is relatively large.) This distribution of voters, like the flipping of many coins, follows a binomial distribution, with N trials and a .5 probability of "success" (voting for a specific candidate) in each trial. According to this distribution, the probability of a specific number k of successes if there is a probability p of success in any one trial is given by

$$\binom{N}{k}p^k(1-p)^{N-k}$$

Plugging in the specifics from our voting scenario gives our answer:

$$\binom{N}{N/2}(1/2)^{N/2}(1/2)^{N/2}$$

which simplifies a little to

$$\binom{N}{N/2}\frac{1}{2^N}$$

We can stop there and be completely correct, but it's still a little hard to understand what happens to your chances of turning the election as the number of voters grows simply by looking at that formula. To see that outcome more clearly, consider Stirling's approximation. Named after James Stirling, an eighteenth-century Scottish mathematician, this method can turn factorials, which show up in our preceding equation's "choose" function, into expressions more readily understood with little loss of accuracy. That method delivers this close approximation of your chances of deciding the election:

$$\sqrt{\frac{2}{N\pi}}$$

For 100,000 voters, say, your vote decides the election with a probability of .0025. Democracy! I'm saying there's a chance—so get out there and vote!

WHICH GEYSER
ERUPTS FIRST?

You arrive at the beautiful Three Geysers National Park. You read a placard explaining that the three eponymous geysers—creatively named A, B, and C—will erupt at intervals of precisely two hours, four hours, and six hours, respectively. However, you just got there, so you have no idea how the three eruptions are staggered. Assuming they each started erupting at some independently random point in history, what are the probabilities that A, B, and C, respectively, will be the first to erupt after your arrival?

Submitted by Brian Galebach

SOLUTION:

Because geyser A erupts precisely every two hours, you know that it must erupt within the first two hours following your arrival. Because geyser B erupts every four hours and because you have no idea when it last erupted, the probability that B will erupt within the first two hours is 1/2. And, similarly, the probability that C, which erupts every six hours, will erupt within the first two hours is 1/3. (In other words, from your point of view, the eruption times for A, B, and C are continuously and uniformly distributed between the time you arrived and two, four, and six hours, respectively, from that time.) There are the following four cases to consider:

1. A, B, and C all erupt within two hours.
2. A and B, but not C, erupt within two hours.
3. A and C, but not B, erupt within two hours.
4. Only A erupts within two hours.

In all of the four cases, each of the geysers that can erupt within the first two hours will erupt first with equal probability. So, for example, in Case 1, A, B, and C all have a 1/3 chance to erupt first. In Case 4, A is guaranteed ($p(A) = 1$) to erupt first.

So to calculate the probability of A erupting first, sum the probabilities of each case occurring multiplied by the probability of A erupting first in each case.

$$p(A) = \left(\frac{1}{2} \cdot \frac{1}{3} \cdot \frac{1}{3}\right) + \left(\frac{1}{2} \cdot \frac{2}{3} \cdot \frac{1}{2}\right) + \left(\frac{1}{2} \cdot \frac{1}{3} \cdot \frac{1}{2}\right) + \left(\frac{1}{2} \cdot \frac{2}{3} \cdot 1\right) = \frac{23}{36}$$

Likewise, we can calculate the probabilities for B and C using the same method. For B, we need only consider Cases 1 and 2, and for C, we only need consider Cases 1 and 3.

$$p(B) = \left(\frac{1}{2} \cdot \frac{1}{3} \cdot \frac{1}{3}\right) + \left(\frac{1}{2} \cdot \frac{2}{3} \cdot \frac{1}{2}\right) = \frac{8}{36}$$

$$p(C) = \left(\frac{1}{2} \cdot \frac{1}{3} \cdot \frac{1}{3}\right) + \left(\frac{1}{2} \cdot \frac{1}{3} \cdot \frac{1}{2}\right) = \frac{5}{36}$$

Probabilities for A, B, and C are 23/36, 8/36, and 5/36, respectively.

WILL THIS SPECIES SURVIVE?

At the beginning of time, there is a single microorganism, the only member of its species. Each day, it and any future members of its species either split into two copies of themselves or die. If the probability of multiplication is p, what are the chances that this species goes extinct?

Submitted by Thierry Zell

SOLUTION:

If $p \leq \frac{1}{2}$, the species will go extinct for sure. If $p > \frac{1}{2}$, it's got a fighting chance. Specifically, it will go extinct with probability $(1/p) - 1$.

Why? Call the probability that the species goes extinct q. That's the probability that the original microorganism dies $(1 - p)$ plus the conditional probability that, if it multiplies, both of its "children" go extinct. Therefore,

$$q = (1 - p) + pq^2$$

Solving for q, we get two solutions: $q = (1/p) - 1$ and $q = 1$. The former is a well-defined probability (i.e., it's between 0 and 1) only as long as $p > \frac{1}{2}$. If $p \leq \frac{1}{2}$, we know the probability of extinction is 1—extinction is guaranteed.

Here's a chart of the probability of extinction given the different probabilities of a single microorganism successfully multiplying. As the probability of multiplication goes up, the probability of extinction quickly goes down:

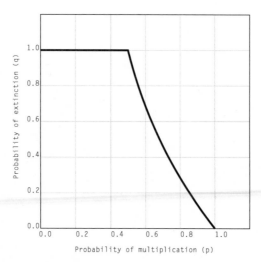

Good luck out there, little buddy.

WILL THE BABY
WALK AWAY?

Your baby is learning to walk. The baby begins by holding onto the couch. Whenever she is next to the couch, there is a 25 percent chance that she will take a step forward and a 75 percent chance that she will stay clutching the couch. If the baby is one or more steps away from the couch, there's a 25 percent chance that she will take a step forward, a 25 percent chance she'll stay in place, and a 50 percent chance she'll take one step back toward the couch.

In the long run, what percentage of the time does the baby choose to clutch the couch?

Submitted by Steve Simon

SOLUTION:

The baby will clutch the couch 50 percent of the time. Let's set up a system of equations based on what we know about the baby's behavior. Let a be the long-run probability that the baby is clutching the couch. Let b be the probability she is one step away, c be the probability she's two steps away, and so on. The baby can reach state a from either state a (staying put at the couch) or from state b (by taking a step backward). So we can define a as $a = .75a + .5b$. Simplifying gives $a = 2b$. The baby can reach state b from state a (taking a step forward), state b (staying put), or state c (taking a step backward). So we can define b as $b = .25a + .25b + .5c$. Simplifying gives $b = 2c$. We can keep going: $c = 2d$, $d = 2e$, $e = 2f$, and so on. Since the baby is always going to be somewhere, we know all these probabilities have to add up to 1. Or to put it another way:

$$1 = a + b + c + d + \dots$$

So, given what we've solved for already, we know

$$1 = a + a/2 + a/4 + a/8 + \dots$$

The right-hand side is a well-known geometric series that converges, because its terms are getting smaller and smaller sufficiently quickly, to $2a$. Therefore, $a = 1/2$.

IS THE BATHROOM OCCUPIED?

There is a bathroom in your office building that has only one toilet. Since only one person can use the bathroom at a time, there is a small sign stuck to the outside of the door that you can slide from "Vacant" to "Occupied" so that no one else will try the door handle when you are inside. Unfortunately, people often forget to slide the sign to "Occupied" when entering, and they often forget to slide it to "Vacant" when exiting.

Assume that 1/3 of bathroom users don't notice the sign upon entering or exiting. Therefore, whatever the sign reads before their visit, it still reads the same thing during and after their visit. Another 1/3 of the users do notice the sign upon entering and make sure that it says "Occupied" as they enter. However, they forget to slide it to "Vacant" when they exit. The remaining 1/3 of the users are very conscientious: They make sure the sign reads "Occupied" when they enter and then slide it to "Vacant" when they exit. Finally, assume that the bathroom is occupied exactly half of the time, all day every day.

Two questions about this workplace situation:

1. If the sign on the bathroom door reads "Occupied," what is the probability that the bathroom actually is occupied?

2. If the sign reads "Vacant," what is the probability that the bathroom actually is vacant?

Submitted by Dave Moran

SOLUTION:

Call the three types of users conscious (C), semiconscious (S), and unconscious (U). C makes sure the sign is "Occupied" upon entering and "Vacant" upon exiting. S makes sure the sign is "Occupied" upon entering but leaves it at "Occupied" upon exiting. U doesn't notice the sign at all.

If the sign reads "Occupied," there are four possibilities:

1. Bathroom is occupied by a C.
2. Bathroom is occupied by an S.
3. Bathroom is occupied by a U, and the most recent previous user who was not a U was an S.
4. Bathroom is vacant, and the most recent previous user who was not a U was an S.

Possibilities 3 and 4 require a bit of explanation. If the bathroom is currently occupied by a U, it doesn't matter how many of the preceding users were also U. All that matters is whether the most recent non-U was an S or a C. If it was a C, the sign would read "Vacant." If it was an S, it would read "Occupied." The same reasoning applies to the fourth possibility if the bathroom is currently vacant.

The probability that the bathroom really is occupied is simply: $[(1) + (2) + (3)]/[(1) + (2) + (3) + (4)]$

> The probability of (1) = $1/2 \times 1/3 = 1/6$ (1/2 chance it is occupied and 1/3 chance user is a C)
>
> The probability of (2) = $1/2 \times 1/3 = 1/6$ (1/2 chance it is occupied and 1/3 chance user is an S)
>
> The probability of (3) = $1/2 \times 1/3 \times 1/2 = 1/12$ (1/2 chance it is occupied, 1/3 chance user is a U, and 1/3 chance most recent user who was not a U was an S)

The probability of (4) = 1/2 × 1/2 = 1/4 (1/2 chance it is vacant and 1/2 chance most recent user who was not a U was an S)

So probability = 5/12 divided by 2/3 = 5/8 = 62.5 percent chance the bathroom is occupied if the sign says "Occupied."

If the sign reads "Vacant," there are only two possibilities:

5. Bathroom is vacant, and the most recent previous user who was not a U was a C (because an S would have left the sign at "Occupied").
6. Bathroom is occupied by a U, and the most recent previous user who was not a U was a C.

The probability of (5) = 1/2 × 1/2 = 1/4
The probability of (6) = 1/2 × 1/3 × 1/2 = 1/12

(5)/[(5) + (6)] = 1/4 divided by 1/3 = 3/4 = 75 percent chance the bathroom is vacant if the sign says "Vacant."

For completeness, the sign will say "Occupied" 2/3 of the time [(1) + (2) + (3) + (4)] and will say "Vacant" 1/3 of the time [(5) + (6)] even though the bathroom is, in fact, occupied half of the time.

Knock knock.

THE MATHEMATICS OF AIRPLANE SAFETY

Riddler Nation is building its first airport and christening its first national fleet of planes. Safety is paramount, and the engineers in charge of the project come to you, the minister of statistics, for some crucial advice. They have two questions:

1. If a four-engine aircraft crashes during takeoff when at least three of its engines fail whereas a two-engine aircraft crashes during takeoff when both its engines fail, is a four-engine aircraft always safer?
2. What must the probability of an engine failure be to make a four-engine aircraft safer than a two-engine aircraft?

Submitted by Philip Schall

SOLUTION:

Here's what you should report back to the engineers:

1. No, not necessarily! This somewhat counterintuitive result can be proven with a simple counterexample. Suppose the probability of a single engine failure is .5. A four-engine plane will crash if three or four engines fail and a two-engine aircraft will crash if both engines fail. So, what are the probabilities of crashes in these cases?

   The probability that the four-engine plane crashes is given by the number of ways three engines could fail times the probability three engines fail, plus the number of ways four engines could fail times the probability four engines fail. Mathematically, that case looks like this:

$$P_4^{\text{failure}} = \binom{4}{3}.5^4 + \binom{4}{4}.5^4 = .3125$$

   Similarly, the probability that the two-engine plane crashes looks like this:

$$P_2^{\text{failure}} = \binom{2}{2}.5^2 = .25$$

   The four-engine plane in this case is actually more prone to disaster.

2. We are looking for those cases when the four-engine plane is safer—that is, those cases when $P_4^{\text{failure}} \leq P_2^{\text{failure}}$. Instead of .5, let's just call the probability of a given engine failing k.

$$\binom{4}{3}k^3(1-k) + \binom{4}{4}k^4 \leq \binom{2}{2}k^2$$

$$4k^3 - 3k^4 \leq k^2$$

$$k \leq 1/3$$

   So whenever the probability of an individual engine failing is less than one-third, the engineers should opt for the larger, four-engine plane.

YOU ARE AN EVIL, DISEMBODIED BRAIN

You are an evil, disembodied brain. You and your two cohorts (for a total of three evil, disembodied brains) get caught doing evil, illegal brain things and are arrested. But your captors, craving entertainment, offer you a game for your release.

Each of you floats in a vat of liquid. On top of each vat, your captors place a hat. The hats are red and blue. The captors determine the hat colors randomly with equal probability. You can see the caps on the other two brains but not on your own. Precisely 10 seconds after the caps are placed, the probes your captors stuck in you and your cohorts will force you all to answer what color your cap is. You can do one of three things:

1. Say "red."
2. Say "blue."
3. Say "pass."

In order to be released, at least one brain must answer with a color, and all the brains that answer with a color must be correct. Given that you are brains, you can assume that you and your cohorts are perfect logicians. However, at no point are you allowed to communicate about a strategy.

What is the optimal strategy? What are the chances you'll be released?

Submitted by Tyler Barron

SOLUTION:

There are two possible types of arrangements of caps. First, all the brains could have the same color cap: red-red-red or blue-blue-blue. Each of those possibilities happens one-eighth of the time, so one of them happens one-fourth of the time. Second, not all the brains could have the same color cap: red-red-blue or blue-blue-red. Those possibilities happen three-fourths of the time.

You should try to exploit those latter possibilities.

If you see two caps of the same color on your cohorts (e.g., red-red), you should guess the other color (blue). If you don't see two caps of the same color, you should say "pass."

This strategy guarantees your release 75 percent of the time. If all the hats are the same color, you're out of luck: You'll all guess incorrectly. But if all the hats are not the same color, one brain will be right and the other two will keep quiet.

SHOULD YOU SHOOT
FREE THROWS UNDERHAND?

The underhand "granny shot" is a much maligned basketball free throw technique. But does it work? Its proponents claim that it improves accuracy because there are fewer moving parts—the elbows and wrists are held with more stability, for example, and the move is symmetric because one's arms are, more or less, of equal length. Let's find out with math how effective the granny shot really is.

Consider the following simplified model of free throws. Imagine the rim to be a circle (which we'll call C) that has a radius of 1 and is centered at the origin (0, 0). The goal is to put the "shot" in the circle. Every time you take a shot it lands at point V, a random point in the plane, with coordinates X and Y. Coordinates X and Y are independent normal random variables, with means equal to 0 and with equal variance. In other words, your average shot is right down the middle, but you vary from perfect the same way left to right and front to back. Finally, suppose that the variance is chosen such that the probability that V is in C is exactly 75 percent (roughly the NBA free throw average).

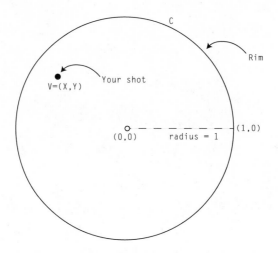

But suppose you switch it up and go "granny style," which in this universe eliminates any possible left-right error in your free throws. What's the probability you make your shot now? (In other words, calculate the probability that |Y| < 1.)

Submitted by Po-Shen Loh

SOLUTION:

The percentage of time you'll make a "granny-style" free throw is about 90.4 percent.

Here's a trick to make things simpler: Instead of fixing the radius of the circle, fix the variance of the random variables to 1. So instead of solving for the variances that would deliver 75 percent accuracy, solve for the radius of the rim first. It doesn't really matter which you do, and this will make things easier.

This solution requires some specific stats and geometry factoids. The circle equation is $x^2 + y^2 \leq r^2$, where r is our radius. The distribution of $x^2 + y^2$ is a chi-squared distribution with two degrees of freedom. (The chi-squared distribution is a commonly occurring probability distribution that results from adding up the squares of normal random variables, such as x and y.) Therefore, to find the radius, we use what we know about chi-squared distributions and solve the following equation for r: $.75 = 1 - e^{-r^2/2}$. That makes $r \approx 1.665$. Using this value, we can find $p = F(1.665) - F(-1.665)$, where F is the cumulative distribution function of a standard normal variable. This gives the solution of about .904.

In other words, if you're a 75 percent free throw shooter using the "normal" method, you can expect to improve to above 90 percent by shooting underhand!

WHO STEALS THE MOST
IN A TOWN FULL
OF THIEVES?

A town of 1,000 households has a strange law intended to prevent wealth-hoarding. On January 1 of every year, each household robs one other household, selected at random, moving all of that household's money into its own household. The order in which the robberies take place is also random and is determined by a lottery. (Note that if Household A robs Household B first, and then C robs A, the households of A and B would each be empty and C would have acquired the resources of both A and B.)

Two questions about this fateful day:

1. What is the probability that a household is not robbed over the course of the day?
2. Suppose that every household has the same amount of cash to begin with—say, $100. Which position in the lottery has the most expected cash at the end of the day, and what is that amount?

Submitted by Max Weinreich

SOLUTION:

First, the probability that our household is never robbed is equal to the probability that each other household does not rob our household. Each robber has 999 potential households to rob (no robber is going to rob himself). The probability that a given other household doesn't rob us is 998/999—there are 999 options and 998 of them aren't us. There are 999 other robbers we have to worry about, so we have to multiply that first calculation by itself 999 times. So, the probability of not being robbed is $(998/999)^{999}$, about 37 percent.

(If the town grew with the number of households increasing toward infinity, this number would approach $1/e$, where e is Euler's number.)

Second, the best position in the lottery is last. The household that robs last ends up with about $137 on average.

It would seem that he who robs last robs best. No one can rob the final household after it has had a chance to go on its robbing spree. But how can we confirm this intuition with nothing but pen, paper, and probability theory? For each household, one of three things will happen.

1. It never gets robbed.
2. It gets robbed after its turn.
3. It gets robbed before its turn but not after.

We've already calculated the probability that No. 1 happens—it's about 37 percent for any given household.

The probability that No. 2 happens, for a household that goes nth in the robbing order, is given by $P_2 = 1 - (998/999)^{1,000-n}$. This is similar to our calculation above. The probability $(998/999)^{1,000-n}$ is the chance of dodging being robbed by all those households after you, and we're subtracting it from 1 to get the chance you are robbed.

The probability of No. 3 happening for the same household is, similarly, $P_3 = (1 - (998/999)^{n-1}) \cdot (998/999)^{1,000-n}$.

If No. 1 happens, the household can expect to wind up with $200: its original $100 plus the average amount left in all the other households it might rob, also $100. If No. 2 happens, the household winds up with $0: It has been robbed and its opportunity to rob has passed. If No. 3 happens, the household can expect to wind up with about $100.10: the $0 it had after it was robbed plus the average of $100,000/999 remaining in the households it might rob.

To arrive at an overall expected value for a given household at the end of the night, we can multiply those probabilities by those expected values and add them all up. This gives

$$((998/999)^{999} \cdot 200) + ((1 - (998/999)^{n-1}) \cdot (998/999)^{1,000-n} \cdot 100.10)$$

This expression is increasing in n—that is, as n gets bigger, the dollar amount gets bigger—so you want to be as late in the lottery as possible. Plugging in 1,000 for n gives an expected amount of $136.83.

HOW LONG WILL YOUR SMARTPHONE KEEP YOU AWAY FROM YOUR FAMILY?

You have just finished unwrapping your holiday presents. You and your sister got brand-new smartphones and opened them at the same moment. You immediately both start doing important tasks on the internet, and each task takes from one to five minutes. (All tasks take exactly one, two, three, four, or five minutes, each with equal probability). After each task, you have a brief moment of clarity. During these moments, you remember that you and your sister are supposed to join the rest of the family for dinner and that you promised each other you'd arrive together. You ask if your sister is ready to eat, but if she is still in the middle of a task, she asks for time to finish it. In that case, you now have time to kill, so you start a new task (again, it will take one, two, three, four, or five minutes, exactly, each with equal probability). If she asks you if it's time for dinner while you're still busy, you ask for time to finish up and she starts a new task, and so on.

From the moment you first open your gifts, how long on average does it take for both of you to be between tasks at the same time so you can finally go to dinner? (You can assume the "moments of clarity" are so brief as to take no measurable time at all.)

Submitted by Olivia Walch

SOLUTION:

One useful way to see the math here is to consider a more general version of the problem, one with the same rules but fewer details. Imagine you and your sister are always on your phones doing one task after another, each of which takes some quantity of minutes, anywhere from one to . . . well, let's just call this number n. While you do these tasks, there is no talking, no other activities—in other words, a normal Thursday night. You still have instantaneous moments of clarity between tasks, but you don't have a special holiday dinner to go to, so if you happen to sync up your own moment of clarity with your sister's (which occurs occasionally), nothing happens, and you each just start again. Let's call these synchronized moments of clarity *sync points*.

The answer to the original puzzle is simply the average amount of time between sync points in this more general scenario. To find that out, we need to know the average frequency of sync points, or number of sync points per minute. The probability of a sync point at any given minute is the probability that you are having a moment of clarity at that minute multiplied by the probability that your sister is also having a moment of clarity.

In our general version, the average time between your moments of clarity is $(n + 1)/2$. That's simply the average of all the integer times from 1 to n. Therefore, the probability that a given minute is a moment of clarity for you is the inverse of that average, or $2/(n + 1)$, and the same goes for your sister. (If the average period of moments of clarity is X, the average frequency of them is $1/X$. Put another way, the longer the average time between moments of clarity, the lower the probability that any given minute will be a sync point.) Since they're independent events, we can multiply those two probabilities, for you and your sister, and find that the frequency of sync points is $(2/(n + 1))^2$.

With an average frequency of $(2/(n + 1))^2$ sync points per minute, the average time between them will be the inverse, or $((n + 1)/2)^2$ minutes, and this carries back over to the scenario outlined in the original riddle. The average amount of time you have to wait for $n = 5$ is $((5 + 1)/2)^2 = 9$ minutes.

Personally, a more realistic upper bound for my tasks these days is around $n = 15$ minutes. That means the expected wait time before my sister and I can do anything together is 64 minutes, which feels about right. Sorry, family!

HOW LONG WILL
IT TAKE TO PAINT
THESE BALLS?

You play a game with four balls: One ball is red, one is blue, one is green, and one is yellow. They are placed in a box. You draw a ball out of the box at random and note its color. Without replacing the first ball, you draw a second ball and then paint it to match the color of the first. Replace both balls, and repeat the process. The game ends when all four balls have become the same color. What is the expected number of turns to finish the game?

Submitted by Dan Waterbury

It takes nine turns on average.

At any point in the game, there are five possibilities for the balls' colors when they're sitting unseen in the box: *ABCD, AABC, AABB, AAAB*, and *AAAA*. (We don't care about the specific colors of the balls, only that they all wind up the same color. *A* could mean blue or red or green or yellow, depending on what happens as we play the game. All that matters is which balls match and which don't.) We might draw a ball of the *A* color and paint a ball of the *B* color to match, or we might draw a ball of the *C* color and paint a ball of the *A* color to match, and so on. Our box begins, for example, with each of the balls painted a different color—*ABCD*. After the first turn, regardless of what we draw and paint, they'll become *AABC*.

But how do we capture this problem mathematically? One option is to set up a system of equations. Let a variable represent the expected number of turns to get from a specific pattern of ball colors to our desired state, where all balls are the same color. For example, if the box has an *AAAB* assortment of ball colors and we take one turn, there is a one-half chance we'll wind up where we started, with colors *AAAB*, and a one-fourth chance we'll paint an *A* ball the *B* color, winding up with *AABB*. We can represent that outcome as

$$AAAB = 1 + \left(\frac{1}{2}\right) AAAB + \left(\frac{1}{4}\right) AABB$$

If our balls' colors are *AABB* and we take one turn, there is a one-third chance the colors will remain *AABB* and a two-thirds chance they'll wind up *AAAB*, which we can represent as

$$AABB = 1 + \left(\frac{1}{3}\right) AABB + \left(\frac{2}{3}\right) AAAB$$

We can do this for the other two possibilities as well:

$$AABC = 1 + \left(\frac{1}{2}\right) AABC + \left(\frac{1}{6}\right) AABB + \left(\frac{1}{3}\right) AAAB$$

$$ABCD = 1 + AABC$$

All we do now is solve this system for the state at the beginning of the game, $ABCD$, using basic algebra or, as my calculus teacher used to say, plugging and chugging. That gives us $ABCD = 9$.

TAKE YOUR VITAMINS

(WITH A SIDE OF MATH)

You take half a vitamin every morning. The vitamins are sold in a bottle of 100 (whole) tablets, so at first you have to cut the tablets in half. Every day you randomly pull one item from the bottle—if it's a whole tablet, then you cut it in half and put the leftover half back in the bottle. If it's a half-tablet, you take the vitamin. You just bought a fresh bottle. How many days, on average, will it be before you pull a half-tablet out of the bottle?

Submitted by Alex Vornsand

SOLUTION:

Including the day you get the half-tablet, it will take about 13.2 days on average.

Let's get our bearings and start day by day. On the first day, there are 100 whole tablets and 0 half-tablets, giving you no chance of pulling out a half-tablet. On the second day, there are 99 whole tablets and 1 half-tablet, giving you a 1/100 chance of pulling out a half-tablet. On the third day there are 98 whole tablets and 2 half-tablets, giving you a (99/100) × (2/100) chance of pulling out a half-tablet. (The 99/100 is the probability you didn't pull out your first half-tablet on the preceding day.) On the fourth day there are 97 whole tablets and 3 half-tablets, giving you a (99/100) × (98/100) × (3/100) chance of pulling out a half-tablet. On the fifth day the chance is (99/100) × (98/100) × (97/100) × (4/100). And so on.

The pattern becomes clear, but now the trick is writing it down mathematically into a single expression we can evaluate. What we need is an expression that, for each day, multiplies the number of the day by the probability that you draw the first half-tablet that day. One option is the following:

$$\sum_{n=2}^{100} \frac{99!}{(99-(n-2))!} \cdot \frac{n(n-1)}{100^{n-1}} \approx 13.2$$

It looks a little messy, but if we start plugging in increasing values for n, we can begin to see how it works and how it matches up with our description of the days above. For $n = 2$ (the first day with any chance of drawing a half-tablet) the expression equals 2 × (1/100), which is the second day times the 1/100 probability discussed above. For $n = 3$ the expression equals 3 × (99/100) × (2/100), which is the third day times the (99/100) × (2/100) probability discussed above. And so on. We've got an expression for the answer!

Simplifying that expression is probably best done with a calculator, unless your vitamins are far more effective than mine.

WILL THE NEUROTIC BASKETBALL PLAYER MAKE HIS NEXT FREE THROW?

A basketball player is in the gym practicing free throws. He makes his first shot and then misses his second. This player tends to get inside his own head a little bit, so this isn't good news. Specifically, the probability he hits any subsequent shot is equal to the overall percentage of shots that he's made thus far. (His neuroses are very exacting.) His coach, who knows his psychological tendency and saw the first two shots, leaves the gym and doesn't see the next 96 shots. The coach returns and sees the player make shot No. 99. What is the probability, from the coach's point of view, that he makes shot No. 100?

Submitted by Mike Donner

SOLUTION:

The correct answer is two-thirds, or 66.666 . . . (repeating) percent. This is an intuitive answer: The coach saw three shots, two of which went in, so it's natural to think that from his point of view the shooter has a 2/3 chance of sinking the last shot. That intuition holds up in this case. Following is a (very rigorous) way to prove it.

Let $H(n)$ be the number of shots hit after n shots have been taken. Given the shooter's specific neurosis, in which the probability he hits any subsequent shot is equal to the overall percentage of shots that he's made thus far, the probability of hitting shot No. 100 is $H(99)/99$ or, because the coach saw him make shot No. 99, $(H(98) + 1)/99$.

To determine the probability distribution of $H(98)$, we can apply Bayes' theorem. Bayes' theorem, named after the eighteenth-century Reverend Thomas Bayes, describes the probability of an event given one's prior knowledge about conditions surrounding the event—the player's other shots made, for example.

Bayes' theorem tells us the probability that an event A happens given that another event B happened. In mathematical notation, that is $P(A|B)$. Bayes' theorem further tells us that probability equals the probability that B happens given that A happened, times the probability of A happening, divided by the probability of B happening. Or in notation

$$P(A|B) = \frac{P(B|A)P(A)}{P(B)}$$

Back to our shooter: The probability that $H(98)$ = some number m (m will range from 1 to 97), given that shot No. 99 was hit, is the probability of hitting shot No. 99 given $H(98) = m$, multiplied by the probability of $H(98) = m$ (in a vacuum), divided by the probability of hitting shot No. 99 (in a vacuum). That's Bayes' theorem in action, which looks like this:

$$P(H(98) = m|\text{No.99 made}) = \frac{P(\text{No. 99 made}|H(98) = m) \cdot p(H(98) = m)}{p(\text{No.99 made})}$$

We can also prove that, for all $N \geq 3$ (a reasonably large number of total shots) and for H and m from 1 to $N - 1$ (the possible number of shots made), $p(H(N) = m) = 1/(N - 1)$. In other words, the probability that the shooter has made m shots in N attempts is 1 divided by the number of attempts minus 1.

For example, for $N = 3$, it's trivial that $p(H(3) = 1) = p(H(3) = 2) = 1/2$. That is, the probabilities of making one out of three and two out of three shots are equal and specifically equal to $1/2$. In other words, if the shooter attempts three shots, he's equally likely to make the third shot and to miss the third shot.

For $N > 3$, the probability of H shots having been hit is the probability of hitting a shot with $H - 1$ on the board, plus missing a shot with H on the board:

$$p(H(N + 1) = m) = m/N \cdot p(H(N) = m - 1) + (1 - (m + 1)/N) \cdot p(H(N) = m)$$
$$= m/N \cdot 1/(N - 1) + 1/(N - 1) - m/(N \cdot (N - 1))$$
$$= 1/N$$

This means that, after shot No. 98, all numbers of total shots made, from 1 to 97, are equally likely, each having probability $1/97$. (One implication is that it doesn't matter how many shots were made when the coach was absent—the right answer would remain the same!)

Because the outcomes of all the shots are equally likely, the probability of hitting shot No. 99 is then $1/2$, or the weighted sum of the probabilities:

$$\sum_{i=1}^{97} (1/97 \cdot (i/98)) = 1/97 \cdot 1/2 \cdot 97 \cdot 98/98 = 1/2$$

Going back to Bayes' theorem one last time, we can calculate that

$$p(H(98) = m | \text{No. 99 made}) = (m/98 \cdot 1/97)/(1/2)$$
$$= 2m/(97 \cdot 98)$$

The probability of hitting shot No. 100 is a weighted sum of the probabilities that H is some given number and that the player will hit the next shot given H.

$$p(\text{No. 100 made}|\text{No. 99 made}) = \sum_{i=1}^{97} 2 \cdot i/(97 \cdot 98) \cdot (i + 1)/99$$

$$= 2/(97 \cdot 98 \cdot 99) \cdot \sum_{i=1}^{97} i^2 + \sum_{i=1}^{97} i$$

$$= 2/(97 \cdot 98 \cdot 99) \cdot (1/6 \cdot 97 \cdot 98 \cdot 195 + 1/2 \cdot 97 \cdot 98)$$

$$= 1/99 \cdot (1/3 \cdot 195 + 1)$$

$$= 2/3$$

WILL THE KNIGHT
RETURN HOME?

Suppose that a knight makes a "random walk" on an infinite chessboard. Specifically, on every turn the knight follows standard chess rules and moves to one of its eight accessible squares, each with probability 1/8.

What is the probability that after the twentieth move the knight is back on its starting square?*

Submitted by Jared Bronski

---

* Knights move in an L-shape: two squares straight and one square over.

SOLUTION:

It's not very likely. The chances are $\dfrac{7,158,206,751,686,848}{8^{20}}$, or about .006.

There are a few ways to get to this result, but none of them are particularly easy. One straightforward but burdensome way is to think of each possible move as a monomial composed of two variables, U and R, where U means "up" and R means "over." For example, U^2R represents the case when the knight moves two squares up and one square to the right. So the U^0R^0 term (the constant term) is the case when the knight has not moved from where it started. With these terms defined, we would then compute the coefficient of the U^0R^0 term in this expression:

$$\left(\frac{UR^2 + U^2R + U^2R^{-1} + UR^{-2} + U^{-1}R^{-2} + U^{-2}R^{-1} + U^{-2}R + U^{-1}R^2}{8}\right)^{20}$$

There are other fancy ways math can get you to the answer as well. Using a Fourier series, you could compute the constant above with this expression:

$$\frac{1}{4\pi^2}\int_0^{2\pi}\int_0^{2\pi}\left(\frac{\cos(x+2y)+\cos(2x+y)+\cos(x-2y)+\cos(2x-y)}{4}\right)^{20} dxdy$$

Joseph Fourier (1768–1830) was a French mathematician best known for developing a series to re-present a mathematical function as the sum of a bunch of sine waves: hence, the cosines in the expression above. If you manipulate and combine the endless ups and downs of sine and cosine curves, you can eventually re-create any other sort of curve that you'd like! Fourier analysis, the branch of math that employs Fourier series, has important applications to processing sound and light waves, which are used in sonar, audio recordings, radio scanners, and elsewhere.

Alternatively, you could tape together a group of chessboards and move a number of knights many, many times. Eventually, you'd arrive at a pretty good approximation of the answer!

CAN YOU STOP THE
ALIEN INVASION?

A guardian constantly patrols a spherical planet, protecting it from alien invaders that threaten its existence. One fateful day the sirens blare: Two hostile aliens have landed at two random locations on the surface of the planet. Each has one piece of a weapon that, if combined with the other piece, will destroy the planet instantly. The two aliens race to meet each other at their midpoint on the surface to assemble the weapon. The guardian, who begins at another random location on the surface, detects the landing positions of both intruders. If she reaches them before they meet, she can stop the attack.

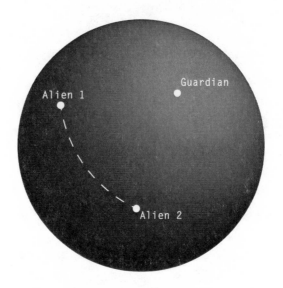

The aliens both move at the same speed. What is the probability that the guardian saves the planet if her linear speed is 20 times that of the aliens?

Submitted by Roberto Linares

SOLUTION:

If the guardian moves at 20 times the speed of the aliens, the planet will be saved roughly 99.27 percent of the time.

The easiest way to think of the solution is to think in terms of the aliens' rendezvous point. The guardian can catch the aliens if and only if she can reach their rendezvous point before they do. If she can get to the rendezvous point first, she can then move directly toward one and catch it. (This simplification relies on the fact that the guardian moves faster than the aliens; a slower guardian would have to use a different strategy.)

If we select two points on a sphere at random, the angular distance θ that separates them has a probability distribution proportional to $\sin(\theta)$. We can call the aliens' initial angular distance from their rendezvous point β. The probability distribution of β is proportional to $\sin(2\beta)$, since they are randomly selected points separated by an angle 2β. Similarly, we can call the guardian's angular distance from the rendezvous point α. The probability distribution of α is proportional to $\sin(\alpha)$.

The overall probability distribution of α and β together is therefore proportional to $P(\alpha, \beta) = k \sin(2\beta)\sin(\alpha)$, and normalizing this to a proper probability distribution, whose integral equals 1, yields $k = 1/2$. (Note that by construction $0 < \alpha < \pi$ and $0 < \beta < \pi/2$.) Finally, the guardian will be able to catch the aliens if her angular distance (α) from the rendezvous point is less than 20 times the aliens' angular distance (β) from the rendezvous point: $\alpha < 20\beta$. Integrating the above probability distribution over this appropriate region of the α–β plane, we get an overall success probability of about 99.27 percent.

It's not too hard to extend this to other speed ratios as well. Noteworthy is the fact that if the guardian travels at the same speed as the aliens, she only has a 1/6 chance of saving the world. If she travels at twice the speed of the aliens, her chances are 50–50. To get a 99 percent success rate, she needs to travel 17.1 times faster than the aliens; to get a 99.9 percent success rate, she needs to travel 54.2 times faster. If she wants to get Six Sigma certified, such that she has a 99.99966 percent success rate, she needs to travel at least 929.4 times faster than the aliens.

GEOMETRY

The entire method consists in the order and arrangement of the things to which the mind's eye must turn so that we can discover some truth.

—RENÉ DESCARTES

CAN YOU WIN
AT TETRIS?

In the video game Tetris, you can, in certain circumstances, completely clear the board after the first five pieces are placed. Knowing that, how many arrangements of Tetris pieces (or tetrominoes) are there that form a solid block that is 2 squares high by 10 squares wide?

Submitted by Bart Wright

SOLUTION:

There are 64 such arrangements. Here is an illustration of all the arrangements:

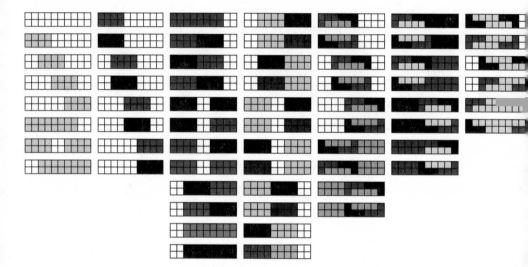

(From light to dark in the diagram above, the pieces are the two-by-two square, the one-by-four line, and then the two L-shaped pieces.)

There is also a lovely pattern. The number of ways to fill a two-by-two, two-by-four, two-by-six, two-by-eight, and so forth, empty space with Tetris pieces is equal to the Fibonacci numbers squared: 1, 4, 9, 25, 64, 169, and so on. Математика—это весело.

CAN YOU BAKE
THE OPTIMAL CAKE?

A mathematician who has a birthday coming up asks his students to make him a cake. He is very particular (he is a constructive set theorist, which explains a lot) and asks his students to make him a three-layer cake that fits under the hollow glass cone that he has on his desk. (The cone was given to him as a prize for proving an obscure theorem long ago.) He requires that the cake fill up the maximum amount of volume possible under the cone and that the layers of the cake have straight vertical sides. (Again, the guy's particular.)

What are the thicknesses of the three layers of the cake in terms of the height of the glass cone? What percentage of the cone's volume does the cake fill?

Here's an example of what the cake could look like, as viewed from the side:

EXTRA CREDIT: What if he had asked for an *N*-layer cake?

Submitted by Jim Crimmins

SOLUTION:

If you make the largest three-tier cake that fits under the mathematician's glass cone, it will fill about 70.2 percent of the cone's volume. If the cone is 1 unit tall, the heights of the tiers, from bottom to top, should be roughly 0.162, 0.182, and 0.219 units.

This is a constrained optimization problem. We want to optimize the volume of the cake subject to the constraints of the dimensions of the cone.

To optimize the volume of the cake, let A_B be the area of the cone's base and H be its height. The volume of the cone is, therefore, $V_{\text{Cone}} = (1/3)A_B H$.

Let the heights of the cake layers be a, b, and c, expressed as percentages of H. The volumes of the three layers, starting with the bottom layer, are then

$$V_a = ((1 - a)^2 A_B) \cdot aH$$
$$V_b = ((1 - a - b)^2 A_B) \cdot bH$$
$$V_c = ((1 - a - b - c)^2 A_B) \cdot cH$$

A layer's additional height (more a, for example) means more cake but has the trade-off of decreasing the radius of that layer's base. As you decrease the radius (width) of the base, you decrease the area of the base by that factor squared, as in πr^2. With each layer's volume we also have to account for the layers that come below it.

The total volume of the cake, V_T, is the sum of those three layers. The percentage of the cone the cake fills—the number we want to maximize—is P.

$$P = V_T / V_{\text{Cone}} = 3(a(1 - a)^2 + b(1 - a - b)^2 + c(1 - a - b - c)^2)$$

We want to maximize P subject to $a + b + c \leq 1$. (The total height of the layers can't exceed the height of the cone, after all.) Solving that—basically not repressing your memories of calculus class and taking partial derivatives and setting them equal to zero—gives us our optimal heights and our optimal volume.

For extra credit, I asked what the cake would look like if it had N layers. The proportion of the cone we can fill as N goes to infinity is clearly 1—our layers become thinner and we can more closely tailor them to the contours of the glass cone. The precise math on the way to infinity gets pretty messy, however, and I'm not aware of a closed-form solution for the optima. But by the time you hit 10 layers, you can fill nearly 90 percent the cone.

Bon appétit!

CAN YOU CON
YOUR SIBLINGS
OUT OF THEIR PIZZA?

You and your two older siblings are sharing two extra-large pizzas and decide to cut them in an unusual way. You overlap the pizzas so that the crust of one touches the center of the other (and vice versa since they are the same size). You then slice both pizzas around the area of overlap. Two of you will each get one of the crescent-shaped pieces, and the third will get both of the football-shaped cutouts.

   Which should you choose to get more pizza: one crescent or two footballs?

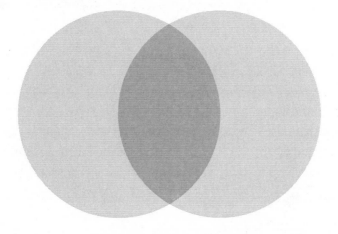

Submitted by Dan Waterbury

SOLUTION:

You'll get more pizza by eating the two footballs.

To show why, begin with the following shape in the middle of the pizzas:

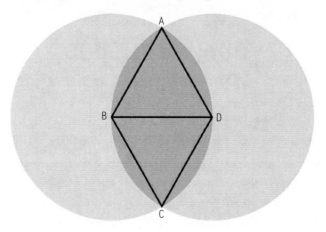

Since sides *AB*, *BC*, *CD*, *DA*, and *BD* are all radii of one of the circular pizzas, they form two equilateral triangles: *ABD* and *CDB*. Thus, the angles *ABC* and *ADC* are each 120 degrees. Therefore, the slice of the pizza on the left bound by *ABC* and the slice of the pizza on the right bound by *ADC* have the area $\left(\dfrac{1}{3}\right)\pi r^2$—they're just 1/3 of each pizza. Let's remove these 1/3 slices from each of the two pieces of football slices you have. You have 2/3 of a pizza. This is your fair share, because there were two pizzas and three eaters. The remaining segments are extra pizza you have earned through mathematics!

Although it wasn't necessary for answering the question, you could also go further and calculate the specific areas of the crescents and the footballs. Assume, for simplicity, that each pizza has a radius of 1. Therefore, two footballs have an area of $\dfrac{4\pi - 3\sqrt{3}}{3}$, or about 2.46; one crescent has an area of $\dfrac{2\pi + 3\sqrt{3}}{6}$, or about 1.91.

Mmm, pizza pi.

SQUARE
THE SQUARE

You are handed a piece of paper containing the 13-by-13 square shown below, and you must divide it into some smaller square pieces. If you are only allowed to cut along the lines, what is the smallest number of squares into which you can divide this larger square? (You could, for example, divide it into one 12-by-12 square and twenty-five 1-by-1 squares for a total of 26 squares, but you can do much better.) Remember, rectangles aren't enough—the subdivided pieces must all be perfect squares.

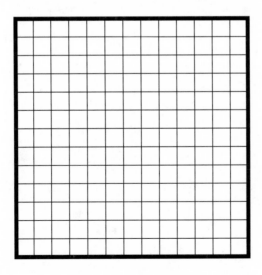

While it's not too hard to find a group of 12 squares that does the tiling trick, the smallest number of squares is actually 11. Here's how it looks:

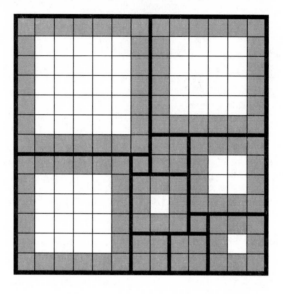

More generally, this problem is known as Mrs. Perkins's Quilt, and it has been the subject of some serious mathematical study. Here are the quilting solutions for some initial squares of different sizes:

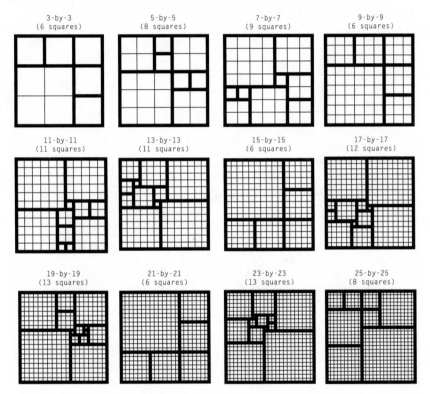

3-by-3
(6 squares)

5-by-5
(8 squares)

7-by-7
(9 squares)

9-by-9
(6 squares)

11-by-11
(11 squares)

13-by-13
(11 squares)

15-by-15
(6 squares)

17-by-17
(12 squares)

19-by-19
(13 squares)

21-by-21
(6 squares)

23-by-23
(13 squares)

25-by-25
(8 squares)

However, while the solutions for larger squares are known for many of these "quilts," mathematicians know of no general solution for a square of side length n. Only recently were 18-square quilts discovered for 88-by-88, 89-by-89, and 90-by-90 quilts, for example. It is known that, for square grids of side length n, the upper bound is on the order of the natural logarithm of n—but not much more than that is known.

Get to work, Riddler Nation!

HOW BIG IS THE
MISSING DICE SLICE?

Fans of Dungeons & Dragons will have fond feelings for four-sided dice, which are shaped like regular tetrahedrons. Some might have noticed, in those long hours of fantasy battle, that if you touch five of these pyramids face-to-face-to-face, they come agonizingly close to forming a closed pentagon. Alas, there remains a tiny angle of empty space left between two of the pyramids.

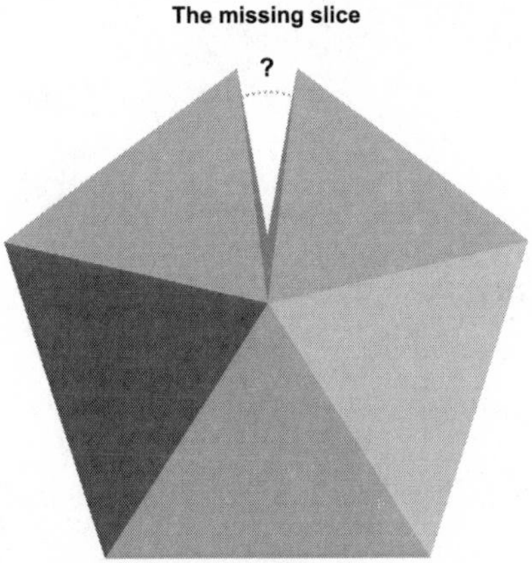

The missing slice

?

What is the measure of that angle?

Submitted by Dan Waterbury

It's about 7.4 degrees.

If you knew what to look for, you could pretty quickly work out the answer to this problem by dusting off and cracking open your encyclopedia. One could probably tell you that a tetrahedron's dihedral angle—the angle between two of its intersecting planes—is about 70.528779 degrees. There are five dice, so there are five such angles. There are 360 degrees in all, so $360 - 5 \cdot 70.528779 =$ a gap of 7.356105 degrees.

But if you eschewed outside counsel, or if you just wanted to start from scratch, here's how you could get there. Assume the sides of the triangular faces of the dice have length 1. Now, begin with a single face and find the height of the triangle.

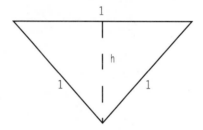

We know from the Pythagorean theorem that $h^2 + .5^2 = 1^2$, so $h = \sqrt{3}/2$.

Now we want to find the angle between two faces—the dihedral angle. Let's call it γ.

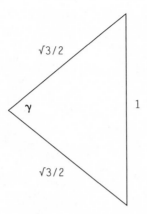

We can find the angle γ using the law of cosines:

$$\cos(\gamma) = \frac{-1 + \dfrac{3}{4} + \dfrac{3}{4}}{2 \cdot \dfrac{3}{4}}$$

Simplifying, $\cos(\gamma) = \dfrac{1}{3}$, so $\gamma \approx 70.53$ degrees. And, again, subtracting five such angles from a total of 360 degrees ($360 - 5 \cdot 70.53$) gives our answer: about 7.4 degrees.

CAN YOU PLEASE THE ARCHITECTURAL ORACLE?

You must build a very specific tower out of four differently colored pieces that can be stacked in any order. But when you start building, you don't know what the correct order is. Upon assembling the pieces in some order, you can consult an architectural oracle who will inform you if zero, one, two, or all four pieces of the tower are in the correct position. Your tower doesn't count as finished until the oracle confirms your solution is correct. How many times should you have to consult the oracle, in the worst case, to assemble the tower correctly?

Submitted by Andrew Simmons

SOLUTION:

In the worst case, you'll need five oracular consultations.

Your first instinct might be that you'd need 24, arguing that the total number of possible tower arrangements—$4 \times 3 \times 2 \times 1$—is also the number of consultations required. But that isn't quite right. You won't need to ask the oracle about every possible order, even in the worst-case scenario, because you learn a lot about the makeup of the correct tower as you go.

Let's walk through one such worst case:

Think of the blocks as lettered and start by stacking them *ABCD*.

1. The oracle says "zero." Swap *A* and *B*: Ask about *BACD*.
2. The oracle says "zero." Therefore, neither *A* nor *B* can be in the first or second position. Swap *B* and *A* for *C* and *D*: Ask about *CDBA*.
3. The oracle says "two." Swap *C* and *D*: Ask about *DCBA*.
4. The oracle says "zero." Swap *C* and *D* for *B* and *A*: Ask about *CDAB*. (You know this is correct but must consult the oracle to confirm, per the rules.)
5. The oracle says "four," and you're done!

We can also illustrate all the ways your construction and consultation could unfold. Each number on the tree represents an answer the oracle might give if your tower isn't quite right. Each step down the tree is an extra guess it may take if you get your first guess wrong.

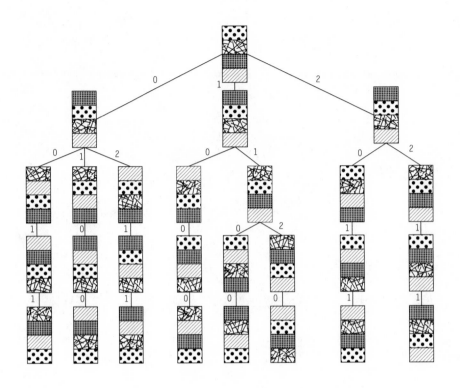

CAN YOU SOLVE THE PUZZLE OF THE ANGRY RAM RUNNING RIGHT AT . . . OH GOD!

A hard-driving sheep farmer is tucked into the southeast corner of a square, fenced-in sheep paddock. There are two gates equidistant from the farmer: one at the southwest corner and one at the northeast corner. An angry ram enters the paddock from the southwest gate and charges directly at the farmer at a constant speed. The farmer runs—obviously!—at a constant speed along the eastern fence toward the northeast gate in an attempt to escape. The ram keeps charging, always directly at the farmer.

How much faster than the farmer does the ram have to run so that the ram catches the farmer just as he reaches the gate?

Submitted by Chris Mills

SOLUTION:

It turns out this particular ram has a golden fleece. If the farmer and the ram begin at different, southern corners of a square paddock and the farmer races to escape at the northern corner nearest to him while the ram chases, always heading directly at the farmer, the ram will catch the farmer only if it's at least 1.618 times as fast as the farmer—the golden ratio!

You can solve the problem with some fancy, elegant calculus. The ram's path is not a straight line or a parabola or part of a circle—rather, it's a pursuit curve. But I like this calculus-free approach: Suppose we fix the speed of the farmer at 1 and a side of the square pen at length 1. The speed we're looking for, the speed of a ram that catches the farmer just as the farmer reaches the exit, is V. Let's solve for V.

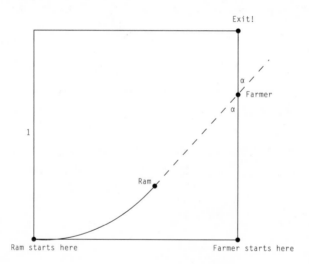

At any given point in the race, the ram's speed in the northerly direction is V times the farmer's speed in the direction away from the ram. That's because the ram moves at V times the speed of the farmer, and the angle between the direction of the ram (the dashed line above) and north is equal to the angle between the direction of the farmer and the direction away from the ram. The angles are both equal to α, as the diagram shows.

Since we know the ram and the farmer will meet exactly at the exit after 1 unit of time, then the average northerly speed of both racers is 1. The ram is moving faster than 1 overall, but its northerly speed matches that of the farmer, who has no east–west vector in his velocity.

The farmer moves away from the ram with an average speed of $1/V$. So the ram must approach the farmer with an average speed of $V - (1/V)$. Since they meet at time 1,

$$1 - \left(V - \frac{1}{V} \right) = 0$$

Solving that gives

$$V = \frac{1 + \sqrt{5}}{2} \approx 1.618$$

HOW BIG A TABLE CAN THE CARPENTER BUILD?

You're on a do-it-yourself kick and want to build a circular dining table that can be split in half so leaves can be added if you need to entertain guests. As luck would have it, you came across a pristine piece of exotic wood on your last trip to the lumber yard. It would be perfect for the circular tabletop. Trouble is the piece is rectangular. You are happy to have the leaves fashioned from some inferior piece of wood, but the aesthetics of the table demand that the main circle come from this extraordinary 4-by-8-foot slab. You devise a plan to cut two congruent semicircles from the perfect piece of wood and reassemble them to form the circular top of your table.

What is the radius of the largest possible circular table that you can make?

Submitted by Eric Valpey

SOLUTION:

It's about 2.70545 feet.

There are two ways that are good candidates for cutting semicircles out of rectangles that maximize the semicircles' area. One is shown here:

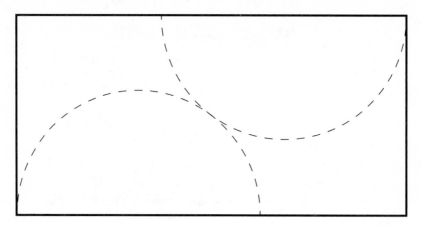

And the other is shown here:

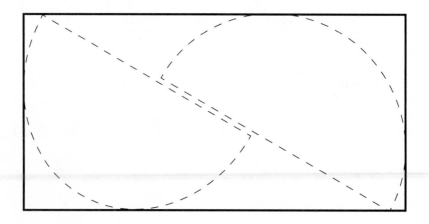

Which of these two patterns produces the larger table turns out to depend on the specific dimensions of our original rectangular piece of wood. If the wooden rectangle is very wide, the first arrangement does the better job as the semicircles can grow to accommodate the wide piece of wood. If the wooden rectangle is narrower, the second arrangement does the better job, leaving less waste. The two do equally well when the ratio of the rectangle's width to height is about 2.5.

So for our 4-by-8 example the second arrangement wins. Let's do a little math to figure out just how big our new table's radius will be. We can break that second arrangement down geometrically like this:

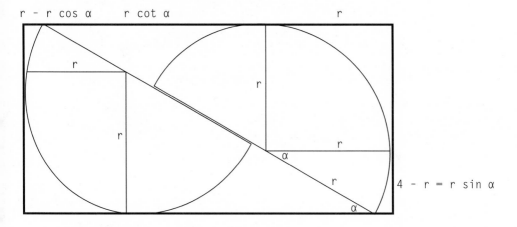

As shown in the diagram above, we can divide the top (the 8-foot side) into three sections of length $r - r \cos \alpha$, $r \cot \alpha$, and r. (Remember SOHCAH-TOA?) After that, it's just a matter of algebra:

$$8 = r - r \cos \alpha + r \cot \alpha + r$$
$$8 - 2r = -r \cos \alpha + r \cot \alpha = 2r \sin \alpha$$
$$\cot \alpha = 2 \sin \alpha + \cos \alpha$$
$$\alpha \approx 0.498945$$

We also know

$$r = 4/(1 + \sin \alpha)$$

So $r \approx 2.70545$.

HOW MUCH WILL
THE PICKY EATER EAT?

Every morning, before heading to work, you make a sandwich for lunch using perfectly square bread. But you hate the crust so much that you'll only eat the portion of the sandwich that is closer to the center of the bread than to the edges so that you don't run the risk of accidentally biting down on that charred, stiff perimeter. How much of the sandwich will you eat?

EXTRA CREDIT: What if the bread were another shape—triangular, hexagonal, octagonal, and so on? What's the most efficient bread shape for a crust-hater like you?

Submitted by @hatathi

SOLUTION:

If you eat only the portion of the square sandwich closer to the center than to the edges, you will eat $(4\sqrt{2} - 5)/3$, or about 21.9 percent, of the sandwich. Very wasteful, your neurosis.

Why? If the sandwich is the unit square (i.e., 1-by-1) and mathematically defined by the area bounded by $|x| < 1$, $|y| < 1$, then the edible portion is characterized by those points that are closer to the center than to the edges. The distance from the sandwich's center can be calculated with the distance equation. In this case, it's essentially the same as the Pythagorean theorem: The distance from the center is the length of a "hypotenuse" from the center to any given point (x, y) on the sandwich, which turns out to be $\sqrt{x^2 + y^2}$. The distance from the edge could be the distance from the left-right edges or the top-bottom edges—it depends on which point of the sandwich we're considering. So we take the minimum of one minus the absolute value of either of our sandwich points' coordinates. That all comes together to look like this:

$$\sqrt{x^2 + y^2} < \min(1 - |x|, 1 - |y|)$$

Plotting that expression, we see that we should expect an answer of a little less than 1/4. The edible section looks like this:

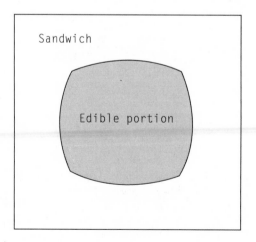

Let's cut out just the top diagonal quadrant of the graph, where $0 < |x| < y$; the rest will immediately follow from symmetry. In this case, the equation simplifies to

$$\sqrt{x^2 + y^2} < 1 - y$$

Solving for y, we obtain

$$y < \frac{-x^2}{2} + \frac{1}{2}$$

showing that the curves that form the edges of that figure are parabolas!

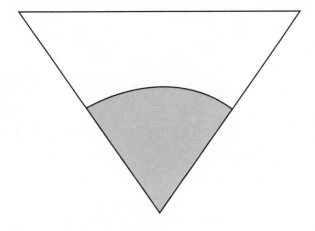

The top parabola terminates at the boundaries of the quadrant where $y = |x|$. Substituting and solving will place the endpoints at $(\sqrt{2} - 1, \sqrt{2} - 1)$ and $(1 - \sqrt{2}, \sqrt{2} - 1)$. Integrating, we obtain the area enclosed between the parabola and the x-axis:

$$\int_{1-\sqrt{2}}^{\sqrt{2}-1}\left(\frac{-x^2}{2} + \frac{1}{2}\right) dx = \frac{2}{3}(2 - \sqrt{2})$$

However, some of that area falls outside our quadrant, specifically two right triangles each with a width and height of $\sqrt{2} - 1$, shown below.

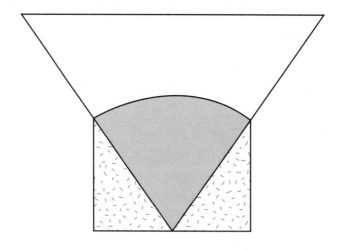

Subtracting these out gives

$$\frac{2}{3}(2 - \sqrt{2}) - (\sqrt{2} - 1)^2 = \frac{4\sqrt{2} - 5}{3}$$

or approximately .219. Since the portion of the bread occupying the top quadrant has an area of 1 (the square has a total area of 4, so each quadrant has an area of 1), this is our final answer.

Who knew eating a sandwich was so complicated? I offered extra credit to those exploring other sandwich shapes. As the number of sides increases to infinity, the bread more and more closely resembles a circle, and the portion you can eat approaches 1/4, or 25 percent. If you're a picky eater, seek circular bread. Maybe create a line of crusty pita breads. Or invest in crumpets. Crumpets are tasty.

THE PRAGMATIC PAPA
AND HIS
FENCED-IN FARM

A farmer has three daughters. He is getting old and decides to split his 1-mile-by-1-mile farm equally among his daughters using fencing. What is the shortest length of fence he needs to divide his square farm into three sections of equal area?

Submitted by Dan Calistrate

As you begin to plan to build a fence, you might sketch out a few blueprints and measure how much fence you need.

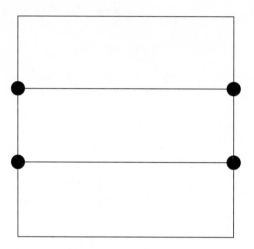

You start off planning to build two east-west fences across the farm. These divide the farm into three identically shaped horizontal slices, giving you the equal areas you desire. This first arrangement uses 2 miles of fence.

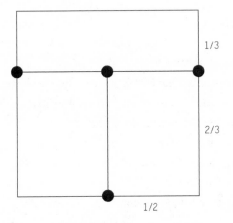

But nothing is forcing you to go east to west. So you draw a blueprint in which a vertical fence line meets a horizontal fence line, as shown in the second diagram. Again, you've divided the farm into three regions of equal area, although this time not identically shaped. This arrangement uses 1 + 2/3 or about 1.67 miles of fence. Much better! If you made it this far, you are conscientious and efficient and have done the agricultural sector of Riddler Nation proud.

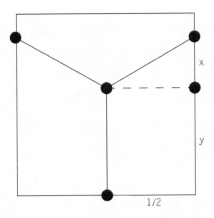

But then you have an epiphany: There's no reason the fences need to be east-west or north-south at all—they could be diagonal! So you sketch out a blueprint like the third diagram in which fences meet in a Y shape to split up the farm. Now things get a little more complicated because you need to figure out where exactly to put the diagonal lines and how much fence this arrangement requires. You'll need a little algebra and a touch of calculus. First, you figure out what the distances x and y must be. You know each region must have an area of 1/3. Using the formulas for the area of a rectangle (base times height) and a triangle (one-half base times height), you see that $y/2 + x/4 = 1/3$. In other words, $y = 2/3 - x/2$.

So how much fence does this arrangement use? It uses one vertical piece (length $2/3 - x/2$) and two diagonal pieces. The diagonal pieces are the hypotenuses of right triangles, so you can get their length from the Pythagorean theorem: $x^2 + (1/2)^2 =$ that length of fence squared. Then take the square root (which is the length of one diagonal) and add two of them to y, which gives the total length of fence in this arrangement. (Call it L.)

$$L = \left(\frac{2}{3} - \frac{x}{2}\right) + 2\sqrt{x^2 + \frac{1}{4}}$$

The farmer's job is to select the x that minimizes the total length L, which can be done by taking the derivative of L with respect to x and setting it equal to zero. That gives $x = \dfrac{1}{2\sqrt{15}}$. Plugging that back into the equation for L, you see this arrangement uses about 1.635 miles of fence. Even better than the second arrangement! If you made it this far, you earned a blue ribbon in the Riddler Nation agricultural expo.

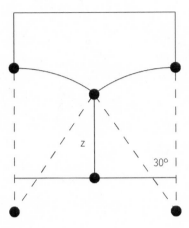

But then you have another epiphany and nearly pass out: There's no reason the fences need to be straight—they could be curved! So you sketch a blueprint like the fourth diagram. It's similar to the third except the "arms" of the fence arrangement are bits of larger circles.

The math here gets more complicated still, and is left as an exercise for the reader. But in the end it turns out that z, the "stem" of the fence arrangement, is about 0.576 mile long, the circle itself has radius 1, the angle pictured is 30 degrees, and the entire length takes about 1.623 miles of fence. The best result yet! Congratulations if you made it this far: You are Riddler Nation's new minister of agriculture.

THE LONELINESS
OF THE LONG-DISTANCE
SWIMMER

Two long-distance swimmers are standing on a beach, right at the water's edge. They begin 100 yards away from one another on the shore, which is a straight line of sand. Both swimmers swim at exactly the same speed. Swimmer *A* heads straight out to sea, directly perpendicular to the shore. At the same time, Swimmer *B* also heads out, swimming exactly in the direction of Swimmer *A* at all times. Over time, Swimmer *B* will approach a position directly in Swimmer *A*'s wake, where he will follow her at a fixed distance.

What is that distance?

Submitted by Scott Cardell

SOLUTION:

The fixed distance is 50 yards.

Why? First, here's an illustration of what the swimmers' paths through the water look like:

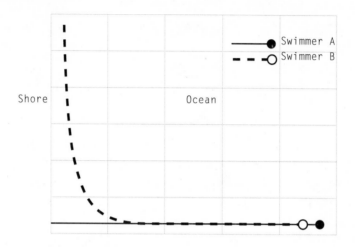

Swimmer *B*'s path toward *A* is not, in fact, a quarter circle, as one might initially assume. Rather, it is a type of "pursuit curve." (See also the puzzle of the angry, fenced-in ram on page 151.)

Let the y-axis be the shore and the x-axis be Swimmer *A*'s path. Have Swimmer *B* start at (0,100). Let d be the distance between the swimmers and let f be the difference in the swimmers' x-axis values. Draw the triangle connecting these three points: Swimmer *A*'s position, Swimmer *B*'s position, and the x-axis point closest to Swimmer *B*.

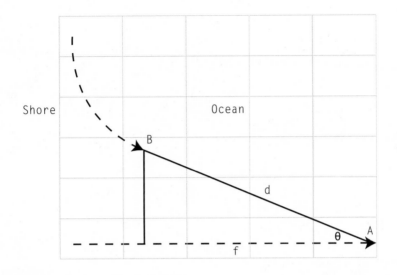

The numbers d and f are the lengths of two of the sides of this triangle. Finally, let θ be the angle between those two sides. Swimmer A's motion is increasing f at the rate v (his velocity) while Swimmer B's motion is reducing f at the rate $v \cdot \cos(\theta)$. (The cosine is the ratio of the length of a triangle's adjacent side to its hypotenuse.) Swimmer B is decreasing d at the rate v while Swimmer A is increasing d at the rate $v \cdot \cos(\theta)$. Thus $f + d$ is changing at the rate $v - v \cdot \cos(\theta) - v + v \cdot \cos(\theta) = 0$. Therefore, $f + d$ is constant. Initially, $d = 100, f = 0$, and $f + d = 100$, so $f + $ d is fixed at 100. When Swimmer B is following directly behind Swimmer A, $f = d$, which means $f = d = 50$ yards.

CAREFUL
WITH THAT
MARTINI GLASS!

It's Friday evening. You've kicked your feet up and have drunk enough of your martini that, when the conical glass is upright, the drink reaches some fraction p of the way up its side. When tipped down on one side, just to the point of overflowing, how far does the drink reach up the opposite side?

Submitted by an anonymous philosophy professor

SOLUTION:

Let's begin with a picture of our cocktail:

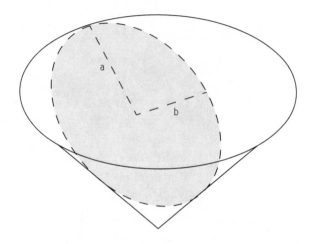

The outline of the liquid's surface, which is the intersection of the cone and the plane of the drink's surface, is an ellipse. So the liquid forms an oblique elliptical cone (a cone whose base is an ellipse and whose apex is off center). But like any cone, whatever the shape of its base and however oblique it might be, it has a volume 1/3 times the area of its base, here πab, where a and b are the semimajor and semiminor axes, times its height h.

Assume your favorite martini glass is a cone with height and radius of 1. We don't lose anything by doing this, because we could just mold any other martini glass into this favorite one in ways that wouldn't affect our answer. Specifically, a right circular cone of any other size and shape can be made from ours by two transformations: vertical scaling to match the shape and three-dimensional scaling to match the size. Both of these affect the volumes of all objects in a constant way, and so the volume of the liquid will still be in the same ratio to that of the cone. And both affect distances along corresponding lines in a constant way, so that the ratio of distances that is the answer to our problem will remain constant.

OK, so now we've chosen our favorite glass. The volume of the upright liquid is $1/3 \cdot \pi p^3$ and so, after expressing that same volume another way, as $1/3 \cdot \pi abh$, we know $abh = p^3$. Let x be the height of the upright liquid (and hence the fraction that is our answer). The area of the triangle in the diagram formed by the liquid is ah, and so

$$ah = \frac{1}{2}\sqrt{2}(\sqrt{2}x) = x$$

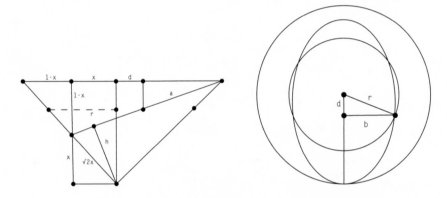

Where r is the radius of the glass cone at the center of the ellipse:

$$r = x + \frac{1-x}{2} = \frac{x+1}{2}$$
$$d = 1 - \frac{x+1}{2} = \frac{1-x}{2}$$
$$b = \sqrt{r^2 - d^2} = \sqrt{x}$$
$$abh = x\sqrt{x} = p^3$$
$$x = p^2$$

Say you've sipped your martini down so far that, when you set the glass on the table, the drink reaches halfway up the side (this is x). If you tip the glass just to the point of spilling, you can expect the drink to reach a quarter of the way up the other side (this is p).

Phew! Time to finish that drink.

WHERE DO THE RANCHERS' CHILDREN ROAM?

Consider four square-shaped ranches arranged in a 2-by-2 pattern, as if part of a larger checkerboard. One family lives on each ranch, and each family builds a small house independently at a random place within the property. Later, as the families in adjacent quadrants become acquainted, they construct four straight-line paths between the houses that go across the boundaries between the ranches. These paths form a quadrilateral circuit path connecting all four houses. This circuit path is also the boundary of the area where the families' children are allowed to roam.

What is the probability that the children are able to travel in a straight line from any allowed place to any other allowed place without leaving the boundaries? (In other words, what is the probability that the quadrilateral is convex?)

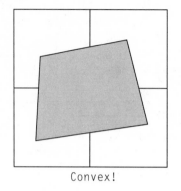

Convex!

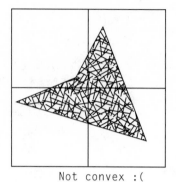

Not convex :(

Submitted by Stephen Carrier

The probability is about 91 percent. The ranchers' children are quite likely to be able to travel in whatever straight lines they please within their allowed area.

You might opt to take a computational, simulation-based approach, randomly building thousands and thousands of houses on thousands and thousands of ranches and checking the convexity of the quadrilaterals they formed. But there is also a precise, analytical answer: $11/6 - 4 \cdot \ln(2)/3 \approx .909137$. It can be calculated with a bit of geometry, probability, algebra, and integration. It takes a smorgasbord of math to analyze a ranch.

Let's begin with this very helpful diagram. Each ranch is represented mathematically on a grid as a 1-unit-by-1-unit square, and the origin—the point (0,0)—is placed at the center of the four ranches.

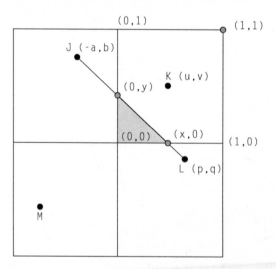

The black points—J, K, L, and M—are hypothetical houses built by the ranching families. For any given arrangement of the houses, at most one house can be the "troublemaker" that makes the quadrilateral nonconvex.

The probability that the troublemaker is K, for example, is the probability that K falls in the shaded triangle in the diagram above. If it fell in that triangle, the children wouldn't be able to walk in a straight line from house J to house L, for example, without leaving their allowed area.

So what is the probability that K falls in that triangle? It's simply the area of the triangle, because the area of the ranch itself is 1. The area of that triangle is one-half its base times its height ($\frac{1}{2}xy$) and the expectation of that area depends on the randomly chosen locations of the neighbors' houses J and L. All that's left now is a bit of careful algebra and calculus.

The expected area of the triangle is obtained by integrating over all the possible randomly chosen coordinates of the two neighboring houses. (For some locations of J and L, x and y will be negative; we can account for that by multiplying our formula by 1/2.)

$$\int_0^1 \int_0^1 \int_0^1 \int_0^1 \frac{1}{2} \cdot \frac{1}{2}\, xy\, da\, db\, dp\, dq$$

Before we can calculate this, we need to know x and y in terms of a, b, p, and q. Knowing the slope-intercept equation for a line, we do a bit of algebra and arrive at $x = \dfrac{bp - aq}{b + q}$ and $y = \dfrac{bp - aq}{a + p}$, which we can plug into our integral above. And because the other three vertices could also cause the problem, we must multiply our probability by 4. Finally, since we're interested in the probability that the quadrilateral is convex, we subtract all this from 1. That gives us an expression for our final answer. (This is tricky to solve by hand!)

$$1 - \int_0^1 \int_0^1 \int_0^1 \int_0^1 \frac{(bp - aq)^2}{(b + q)(a + p)}\, da\, db\, dp\, dq \approx .909137$$

ACKNOWLEDGMENTS

Much like our modern understanding of ancient Egyptian mathematical papyri, this book would not exist but for the hard work and abiding belief of a large set of devoted individuals. As I type each name below, I will pause in gratitude while performing a brief arithmetic calculation in my head in tribute.

Thanks go to my deeply thoughtful editor at FiveThirtyEight, Chadwick Matlin, in collaboration with whom this whole puzzle endeavor was hatched; my editor at W. W. Norton, Tom Mayer, for his vision, patience, and the coffee; Mike Wilson, now of the *Dallas Morning News*, whose decision to give me my start in the writing business is more perplexing than any puzzle in this book; ditto Nate Silver, FiveThirtyEight's indefatigable founder and editor in chief, who was kind enough to bless this project; the endlessly able copy editors Meghan Ashford-Grooms, John Forsyth, Colleen Barry, Sara Ziegler, and Natalia Ruiz, who kept and keep "The Riddler" column stylish each week; the bafflingly gifted visual journalists Gus Wezerek, Ella Koeze, Julia Wolfe, and Rachael Dottle, who routinely improve the column with their charts and illustrations; Dhrumil Mehta, my office neighbor and data wizard; Carl Bialik and Mai Nguyen, who helped check my math; and David Firestone and Stephanie Roos, who already know why.

If Riddler Nation convened a cabinet, the following citizen-readers would be its inaugural members, given their regularly outsize and brilliant contributions to both the puzzles and the solutions in the column and this book: Laurent Lessard, Zach Wissner-Gross, Hector Pefo, Diarmuid Early, Guy Moore, Dan Waterbury, Ian Rhile, Tim Black, Tyler Barron, Rob Shearer, Sean Henderson, Dave Moran, Rajeev Pakalapati, Zack Segel, Lucas Jacobson, Luke Benz, Christopher Long, Jason Ash, Laura Feiveson, Jason Weisman, Sawyer Tabony, Satoru Inoue, Dennis Wolfe, Po-Shen Loh, and others whom I have mistakenly omitted due solely to my own deficiency. A grateful nation thanks you!

Much love to my parents, Phil Roeder and Mary Tabor, in whose house is still maintained a large cardboard box of books from my childhood labeled "Puzzles and Games." I hope this book merits future inclusion.

Much love also to Emily Schmidt who—when asked if and how she would like to be acknowledged for the incalculable support she has given me—simply laughed and fixed a pot of tea.

The Riddler is, above all, a large-scale collaboration. Thank you to everyone in Riddler Nation who has ever sat down to solve a puzzle.